Data Mining for Managers

This page intentionally left blank

Data Mining for Managers

How to Use Data (Big and Small) to Solve Business Challenges

Richard Boire

palgrave
macmillan

DATA MINING FOR MANAGERS

First published in 2014 by
PALGRAVE MACMILLAN®
in the United States—a division of St. Martin's Press LLC,
175 Fifth Avenue, New York, NY 10010.

Where this book is distributed in the UK, Europe and the rest of the world, this is by Palgrave Macmillan, a division of Macmillan Publishers Limited, registered in England, company number 785998, of Houndmills, Basingstoke, Hampshire RG21 6XS.

Palgrave Macmillan is the global academic imprint of the above companies and has companies and representatives throughout the world.

ISBN 978-1-349-48786-8 ISBN 978-1-137-40619-4 (eBook)
DOI 10.1057/9781137406194

Library of Congress Cataloging-in-Publication Data

Boire, Richard.
Data mining for managers : how to use data (big and small) to solve business challenges / Richard Boire.
pages cm

1. Management—Computer programs. 2. Data mining. I. Title.

HD30.2.B6373 2014
658′.056312—dc23 2014015559

A catalogue record of the book is available from the British Library.

Design by Newgen Knowledge Works (P) Ltd., Chennai, India.

First edition: October 2014

10 9 8 7 6 5 4 3 2 1

Contents

Figures

FOREWORD

TERMS AND SUBJECTS SUCH AS BIG DATA, DATA MINING, AND PREDICTIVE analytics are commonly used in today's work place, but are they really understood by the community at large? The emerging field of analytics is growing at a significant pace, but it is not new. Much of the growth is due to our ability, thanks to computer technology, to capture, store, manage, and act on vast amounts data. Data mining and analytics are used in almost every industry and facet of our society. From understanding consumers and what they are likely to purchase to identifying potential terrorist activities, predicting the likelihood of someone being a good employee, and understanding the probability of where a baseball player is likely to hit the ball, data mining and analytics are everywhere—whether you notice it or not.

In today's big data world, dealing with large volumes and varieties of data at rapid speed is possible, but what to do with this data is still a challenge for most organizations. It takes experience and knowledge to navigate the big data maze and convert raw data into a powerful and meaningful resource. In a time when technological capital has outpaced human capital, this book will help accelerate your learning curve.

In Richard Boire's book, *Data Mining for Managers*, you will learn from a leader in the field of analytics. He has been a practitioner in the field of data mining since well before terms like "big data" were conceived. This book will allow you to benefit from his more than 25 years of experience and take advantage of the important and vital lessons he has learned along the way.

Richard's practical and pragmatic approach to solving business problems are on display throughout the book. Leveraging real business examples, you will learn how to evaluate or define the problem to be solved; how to leverage internal and/or external data to solve the problem; how to build, create, and select the best tool or solution; and finally how to properly implement and measure the impact of the solution.

You will also benefit from his experience as an educator and a mentor to many. You will gain insights into what makes analytical initiatives successful. Effective data miners (or data scientists) can't work in isolation. Successful data mining initiatives require team work and interaction with other key disciplines, such as data base managers (or data technicians or IT) and the end business users, often referred to as the "subject matter expert (SME)" or the value architect. And the most successful solutions require a combination of both art and science.

I am proud to say that for the past 25 years I have been a colleague, partner, and friend of the author Richard Boire. Much of what I have learned about data mining and predictive analytics I owe to him. This book is an opportunity for you to share in this learning, and it is a must-read for anyone new or experienced in the area of data mining and analytics.

Larry Filler
Partner,
Boire Filler Group

Acknowledgments

A practitioner-based book like this can only be written on the basis of its author's real-life experiences. In this journey, there are indeed organizations and individuals I need to thank since without them the book would not have been written.

Reader's Digest and American Express were key organizations in my early career, providing the foundation and culture to build my learning in data mining. On a personal level, I would like to thank David Young, vice president of Direct Marketing at American Express Canada and a key mentor, for demonstrating how to socialize the benefits of data mining to key business stakeholders.

As the data mining discipline is still somewhat new and emerging, the knowledge base continues to evolve. It is my belief that a book that could leverage my learning and experiences as a practitioner, would add to the knowledge base of this discipline. Of course, the most important organization in providing these experiences has been the Boire Filler Group. For 15 years, our team has been able to provide the benefits of data mining to organizations in all industry sectors. We owe our success to the valued and talented people as well as the leadership within the Boire Filler Group. No thanks would be complete without recognizing the leadership of this organization under my partner, Larry Filler. He has been instrumental in building our business but more importantly in allowing our organization to provide work in a variety of different sectors, which is the basis of most of the content in this book.

On a more personal note, I would like to thank my family and specifically my wife, Clare. I would like to dedicate this book to her as her undying patience and support have been the real catalysts for my success and enthusiasm in this discipline.

This page intentionally left blank

CHAPTER 1

INTRODUCTION

THE CHALLENGE IN WRITING A BOOK ON DATA MINING IS TO differentiate its content from the plethora of other books, articles, and seminars on the same topic. The explosive growth of data has certainly fuelled the need for more discussion on this topic, yet, minimal content has been actually devoted to the topic of data. The focus of the book is on data and its use in shifting the key levers of your business, and to ultimately understand how this drives ROI. Data mining is highly popular now because most businesses understand the value of information and of using it to make more profitable decisions. In most organizations today, data mining either is or will soon be a key business process. Since data mining is relatively new, knowledge and understanding of it is critical for its successful implementation. Certainly, consulting has grown in this field, and many people profess to be experts in data mining. Companies must realize that this is a new field and that it is difficult to find practitioners with extensive experience. As with any other discipline, the key to becoming knowledgeable and acquiring expertise is to put ideas into practice and observe the results. Although much of this practical knowledge has been acquired in the context of direct marketing, the approach and processes of data mining are similar in other areas. The key to attaining data mining excellence is in how one deals with the data. Data Mining is about data and how we derive knowledge from this information. This is the essence of data mining, as you will realize in the course of reading this book.

THE DIGITAL IMPACT

According to experienced practitioners, one of their greatest challenges today is optimizing the use of data mining, given the explosion of new software and technology currently available. The other significant challenge facing practitioners is the use of data mining in the web environment. For example in the marketing sector, data mining analyses can be conducted from the moment that a campaign is launched. Access to this type of information has already created the demand for tools and technology that will increase both the volume and the speed of analyses. In the past, analyses used to take six to eight weeks to deliver useful information. With web and digital data, we can gather information instantaneously. Due to this increased access to data, the learning environment flourishes where businesses can more quickly derive those meaningful insights and thereby increase their ability to make better decisions.

SERVICES

The field of data mining is still in its infancy, and it will be very interesting to observe the impact it will have on business. Data mining's reliance on technology suggests that the industry is quickly evolving in terms of available tools and software. This development will also place tremendous demands on practitioners. There is a growing demand for data mining practitioners, a demand that is not met in the current market and is not even being addressed by major universities.

ART AND SCIENCE

Although the technology is becoming more sophisticated and complex in terms of providing additional targeting capabilities, there will always be an element of art in building data mining solutions. Understanding the data without any cogent business understanding of the results will generate conclusions and recommendations that can be very misleading.

For instance, through a data mining analysis, one retailer discovered that the sales of beer and diapers were very highly correlated. Does that really mean that customers who buy beer are most likely to also purchase diapers? Upon further investigation, it was found that these results were really due to the fact that beer and diapers were randomly placed next to each other in the store without any prior knowledge of purchase behavior. As it turned out, young fathers were shopping for diapers late at night and decided to pick up some beer along with the diapers. The important lesson here is that further investigation was warranted to truly uncover the "insight" delivered by the raw numbers. In this case, the data mining analysis was skewed simply by product distribution in the store and did not reflect consumers' intentions.

Another example was that tenure or newer customers were suddenly more heavily predisposed towards purchasing products of a certain organization. Upon further investigation, it was revealed that this same company acquired another company with all its customers in the last year and then began promoting its products to the acquired company's customers. The tenure of these customers from the acquired company was reported on the database as being 1 year tenure. Using lower tenure as a characteristic in selecting customers on a going forward basis for this company's products would be flawed.

The art component of data mining also allows the analyst to better interpret results. With a solid understanding of the business, the data miner can better understand why certain results are occurring. At the very least, he or she can better investigate certain facets of the business that may illuminate the underlying reason for the results. Although data mining is number-intensive, knowledge of the business combined with the numbers is what allows the practitioner to deliver an optimal analysis.

Data mining is not solely about technology; it is about the use of technology to arrive at business solutions that optimize return on investment (ROI). This requires that organizations invest in intellectual capital as well as in technology such as software and new database systems. In fact, successful organizations will prioritize their investment on the intellectual side rather than on the technological front. The key to successful data mining is producing solutions that can truly optimize ROI at its most granular level; in most cases, this is at the customer level.

Data mining continues to evolve as a science and as an industry discipline, but to be truly successful as a practitioner of data mining, analysts must realize that there is an element of art involved. At the same time, data mining is now widely seen as contributing huge benefits to our society. For example, the ability to target any product to the right person at the right time lowers marketing costs while often enough yielding incremental revenue. These economics often translate into ROI increases of 100% or more. Ultimately, this increased ROI is about generating more revenue with lower costs. In the credit risk/fraud area, often considered the birthplace of data mining, as we will see in later chapters, analysts attempt to identify individuals who are likely to default on payments or, in the case of fraud, individuals who exhibit unusual patterns of spending. Even an improvement of only 1% in the areas of credit risk and fraud prevention due to data mining can translate into millions of reduced credit card losses that directly impact the bottom line.

INDUSTRY PERSPECTIVE

HEALTH

In hospitals and the health sector in general, health professionals analyze vast amounts of data and information to identify patients' problems or diseases. By having access to the patient's history and also to perhaps hundreds of other patients' histories, health care professionals are better equipped to identify the given challenge or problem even in relatively unique situations. Access to prior patient history that encompasses vast amounts of data can often lead to a unique treatment for a given patient. Thanks to data mining, today's medical professionals can analyze and combine vast amounts of data pertaining to patient histories to more easily provide optimum care and treatment for a given patient.

GOVERNMENT AND LAW ENFORCEMENT

Government agencies use data mining largely as a law enforcement tool. If we think of data mining as a tool for discovering knowledge, we can understand that the key to knowledge discovery is the ability to detect unique patterns in the data. In law enforcement, we all understand the notion of "detective work" as detectives personally deal with their own repositories of data. These repositories of data consist of notes in a notebook, previous experiences, and hunches. The acquisition of this type of knowledge is very time and labor intensive. Based on this knowledge, detectives then attempt to identify the unique pattern that will lead to a potential arrest. With data mining, a higher level of automation can be utilized, and vast amounts of data and information from a variety of detectives can be collected from more than one crime scene. The data is then compiled into one analytical file or database and statistically or mathematically analyzed to uncover unique patterns of behavior or events that could help to solve the crime at hand.

Besides solving a specific crime, using this data may also help to determine which areas in a city are most prone to specific crimes. In data mining, we can architect the data by looking at the information prior to a particular event, which allows us to be predictive in terms of future events. With this type of information, the police can then better allocate resources. This smarter allocation of resources means the city can send both the appropriate number and type of officers based on the type and volume of crime that most often occurs in a given area. For example, New York City did exactly this under Mayor Rudolph Giuliani in an attempt to reduce the horrific crime rates, in particular the homicide rates, that characterized the city in the late 1970s and early 1980s. New York's homicide rate is now down to less than a third of what it was in the late 1970s. The use of data mining technology is one of the key reasons for this significant decrease.

The notion of using data mining as a law enforcement tool has certainly gained even more prominence after 9/11. For example, in response to the terrorist attacks of September 11, the US government established a department to oversee the compilation of data on individuals gathered by various government organizations and the use of that data in combating terrorism. This massive database contains information pertaining to areas such as health, insurance, and banking. Essentially, the government could have a dossier on each individual's activity and behavior over the course of that person's life. Aside from the obvious privacy concerns, one of the arguments against such a project is that the development of tools and solutions for preventing terrorist activity requires a large number of observations. In the case of 9/11 and of perhaps preventing future terrorist attacks the argument is that we would not be able to gather enough data to build profiles of Al-Qaeda terrorists since, we have only 19 data points (19 terrorists) to contrast with information about the remaining 300 million nonterrorists in North America.

Various debates and discussions have centered on the use of data to identify potential terrorists. Among the fallout from these discussions has been the notion of "racial profiling" and the obvious implications of this practice. With information gleaned from data mining, law enforcement organizations can focus on certain sectors of the population that are considered high-risk (i.e., racial profiling).Yet issues as sensitive as racial profiling will need to be addressed by examining how to balance the need for public security with that for protection against law enforcement discrimination. A number of key stakeholders will need to be involved in these thought-provoking debates. Ultimately, I hope that this process will result in a policy that clearly lays out the guidelines regarding the use of racial profiling in the context of law enforcement.

Real-World Experience of Data Mining

When I began my career as a direct marketer in 1983 after graduating with an MBA, I was fortunate enough to apply my academic knowledge in the areas of statistics. Like most young graduates, I was confident that I would make a difference in my new company. Yet although I contributed to the organization in fulfilling the duties of my position, the organization contributed far more to me by providing a basis on which to build my career, namely, the application of statistics in business. I learned to follow the esoteric and arcane principles of statistics but not adhere to them in the strictest sense. For example, in most cases, the traditional assumptions regarding the statistical analysis of sample groups are not followed because most analytical situations using statistics in the business environment deal with nonnormal samples. A normal sample is when half the sample lies below the mean of a certain sample characteristic (i.e., age, income, etc.) and the other half lies above the mean with the distribution of data looking like a bell curve. Yet, businesses continue to apply these techniques, much to the pure mathematician's horror, because they work and produce acceptable results that yield significant incremental benefits. The recognition that it was acceptable to bend the rules of statistics was perhaps my most important introduction to the application of statistics in the world of direct marketing. In the world of academia, we were often used to regression results that produced R^2 of .7 or more. (This topic will be discussed in more detail in later chapters.) In the world of direct marketing, it was not uncommon to observe R^2 results of .05 or less, depending on the data. Statistical results that would be totally unacceptable in the academic statistical arena were commonly applied in the business world. The reason for this divergence in opinion between academia and business is that each side views the success of a given solution in very different ways. We will discuss this in more detail in later chapters.

As with any book on a topic of great significance, it is important to differentiate this book from others. If you are looking for mathematical and technological nuances and insights regarding data mining, this book is not for you. However, if you want a more practical business perspective of data mining, you have found the right book. Using this perspective, I have decided to focus on data mining from a practitioner's viewpoint to allow you to benefit from my more than 30 years of applying a number of tools and techniques to countless business situations. I hope my insights will shed light on what works and what doesn't work under certain conditions. As with any book on data mining, there will be discussion of the technology and mathematics involved, but the focus is on how data mining impacts a business. However, I do not intend to turn readers into data mining specialists. That is, if you already have experience with data mining, this book can offer another point of view to consider. In addition, this book provides insights into data mining in a Canadian context. Most books on data mining are written in the United States from an American perspective. Adding a Canadian perspective to the discussion will enhance overall understanding of the topic. If you find this book serves you as a useful reference regarding specific data mining tactics and their impact on a particular business problem, then I will have achieved my goal.

Enjoy.

CHAPTER 2

GROWTH OF DATA MINING—AN HISTORICAL PERSPECTIVE

IN WRITING A BOOK ON ANY TOPIC, IT IS ALWAYS IMPORTANT TO understand its history. What were the challenges and developments of the past that catapulted data mining to the forefront of business today? To get a sense of data mining's history, it is useful to speak to its early practitioners, namely, the direct marketers of the large major catalogue companies and publishing houses.

As a recent MBA graduate back in 1982, I was very fortunate to be hired by one of these pioneering direct marketing firms, Reader's Digest. At that time, I was hired to work in the company's list selection department as a regression analyst. Although we did develop predictive regression models for direct mail campaigns, we were responsible for all analyses or segmentations dealing with individual-level data on customers. My colleagues and I were the customer knowledge gurus at Reader's Digest and saw the huge ROI payback of this business culture; naturally, I thought that this business process was typical of most organizations. Conversations with my MBA business colleagues who were working at leading institutions in various industries in the early eighties revealed that this type of work was not being done. Back then, we had no fancy titles, such as business intelligence specialist, knowledge discovery analyst, data scientist, or CRM business analyst, because those were the "pioneering" days of data mining and predictive analytics.

I realized the tremendous business potential of data mining in improving overall ROI and saw that Reader's Digest was one of only a handful of organizations doing this kind of work. I realized the tremendous entrepreneurial opportunities in data mining. However, one barrier to becoming an entrepreneur in this field was technology. Computing equipment for capitalizing on these statistical techniques used in direct marketing campaigns came at tremendous costs because in 1982, there were virtually no PCs yet, and all our work was done on mainframes. Yet, the analytical environment at Reader's Digest was very robust and efficient. We had an extensive campaign history for each customer. We knew when customers were promoted to, how often they received promotions, what types of promotion they received, and how often they had received promotions since their last purchase. With this history of promotion and purchase behavior, we were able to quickly develop robust models. We had multiple models for each product line based on customers' specific promotion history. For instance, in the one-shot book (OSB) product line, we had one model for the segment of 0–3 OSB offers since last order, another model for the segment of 4–6 OSB offers since last order, and another model for the segment of 6+ OSB offers since last order. Furthermore, for each product promotion history segment, we had multiple models for specific times of the year in order to take into account the impact of seasonality on campaign performance.

In many cases, our segmentation strategy and schema were developed from past results and learning from prior campaigns. By using our business intuition and judgment on these past results, we were able to make sound business decisions regarding any new business segments. Statistics were used when it was clearly evident that incremental business results would be achieved. Reader's Digest was always proactive in testing out new technologies and approaches in order to be at the forefront in direct marketing. This included being better able to target customers rather than simply merge or purge lists or personalizing names in a direct mail piece.

Statistics were used judiciously; that is, our methods for evaluating different statistical techniques in a live business initiative were not always based on the statistician's purest viewpoint. For instance, in the academic world, when evaluating multiple regression techniques, values of R^2 in the neighborhood of 70% to 80%, where 100% represents perfection, were not unusual. Essentially, this benchmark measures the amount of explained variation produced by the model relative to the total variation of the data (the concept of R^2 will be explained in more detail in a later chapter). Yet, in a direct marketing campaign, it was quite normal for R^2 to range only from 1% to 4%. Even with these depressed R^2 results, many of these campaigns were largely successful as a result of their predictive models. In the business world, large random variation is the norm rather than the exception. By explaining just a small component of this variation, the predictive models yielded significant business benefits for Reader's Digest. The discipline of measuring techniques and their impact on live business data is critical to any evaluation process. As a result of this discipline, Reader's Digest was able to establish a business culture in which statistical learning was always combined with business learning.

As with any new business process, the right culture must be in place if data mining is to produce useful results. In particular, employees must have a positive mind-set and be willing to embrace change and new learning. Hiring the best people and purchasing the best technology will produce less than satisfactory results if the organization itself does not have the right fit for data mining. For example if an organization is starting at ground zero in data mining, the question is how it becomes a "best practices" leader in this discipline.

For marketers, using acronyms and buzzwords when describing the latest ad or campaign comes naturally, and it is the same when they describe a process or culture that changes how they think and execute processes in their marketing work environment. It is certainly the case with data mining. Many books and articles have been written on this topic, and there is no shortage of people claiming to be experts in this field. Given the growing number of seminars on this topic, it would appear that this expertise in this field is growing exponentially. Yet, in reality data mining is still in its infancy.

As with other changes in our society today, technological advancements have created an environment that allows more players to play the game. In the late seventies and early eighties, only the large direct marketing businesses, such as Reader's Digest, practiced data mining. They could do this because they invested millions of dollars in the development of an appropriate marketing database. This investment in technology was justifiable since direct marketing was the core business of those companies.

Direct marketing often gets a bad name as being simply "junk mail." But one of the core principles of successful direct marketing is what data mining is all about. In other words, the ability to uncover knowledge about customers and/or prospects and apply this learning to future business initiatives is essential to the success of any direct marketing or Customer Relationship Management CRM initiative. As practitioners of successful direct marketing, the people at Reader's Digest always knew who to promote to, what to promote, and when to promote. However, for most other organizations, this was not the case because direct marketing represented only a small portion of their marketing budget. The business paradigm for most marketers in the late seventies and early eighties was optimization of revenues, and this was supported by the prevailing business culture of that time. Yet, the technological advances of the past 25 years have not only changed how society functions but have also ultimately changed its expectations. With technology, we are expected to do more with fewer resources.

Consequently, consumers are now bombarded with all kinds of communications and offers in the hope that they will respond appropriately to each company's message. With the web becoming more and more important as a marketing channel, we now have the ultimate medium for collecting and acting on individual information. This means that decisions regarding how we communicate with customers in this medium can be made and changed instantaneously. In the old direct marketing environment, we had to wait several weeks after the launch of a campaign before we received any information from consumers. The dynamic and interactive nature of the web requires that data mining becomes a firmly entrenched business practice.

Increased access to information and, more important, the capability to use it are significant competitive advantages in today's business environment. Organizations operating in this environment can view all their marketing initiatives from the standpoint of ROI. At the same time, these organizations can gain insights into what worked and what didn't. This dynamic learning process allows businesses to always strive for improvement. This is the key to successful data mining. The cycle of campaign execution and learning provides the ongoing feedback for improving future marketing efforts.

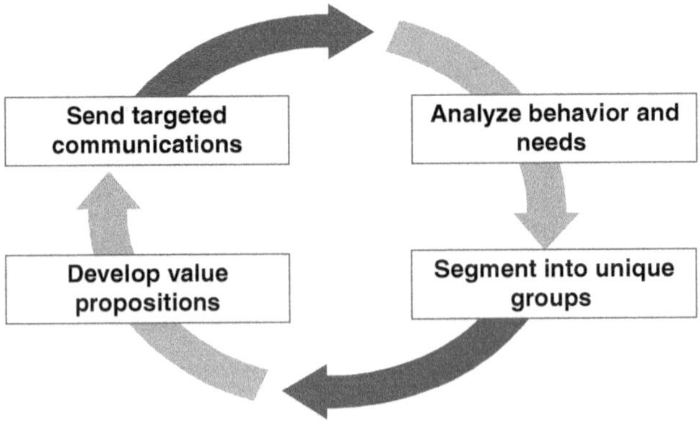

Figure 2.1 The Cycle of Data Mining Learning and Execution

This is best illustrated through figure 2.1. A live case study is presented below which shows how an organization has evolved through its data mining practices.

CASE STUDY: AMERICAN EXPRESS CANADA: HOW DATA MINING PRACTICES EVOLVED OVER TIME

In the 1980s, American Express was striving to increase its membership base. In fact, its share price was directly impacted by how many new cards were acquired (CIF, cards in force). Toward the end of the eighties, CIF targets were being achieved but the cost per new card had doubled. These cost inefficiencies were unacceptable to the organization. In order to resolve this dilemma, the company began to purchase names from lists that seemed to fit the profile of an American Express cardmember. Lists from magazines, such as *Business Week*, *Forbes*, *Fortune*, and others, produced very good acquisition results. However, the number of names on these lists was too small to provide the prospect universe necessary to generate Amex's aggressive CIF targets. Ultimately, Amex needed a very large prospect universe and tools to generate new cards in a cost-efficient manner. The solution was twofold.

First, Amex built a prospect history database that compiled information at the level of individual prospects. Examples of some key information included the promotion history and gender of the prospect. With this database, Amex began to sort its prospects into two key segments based on promotion history. The two key segments were new names and previous names.

The second solution was to then build predictive models that could be applied to both universes in order to optimize response rates. Using overlay aggregate-level data from Statistics Canada, gross response models were built and provided a 50% lift in acquisition performance. However, it was also revealed that approval rates had actually decreased. The company discovered that the best responders were those most likely to seek credit. The learning from this campaign indicated that Amex needed a targeting tool that optimized net response rate (gross response and approval) and not just gross response. A net response model was developed, and once again the company achieved a 50% lift in acquisition performance. However, this time it achieved much better cost-efficiencies regarding the new applications because these applicants were more creditworthy. Ultimately, this led to a decrease in credit losses for this group of customers in the first

12 months of being Amex customers. Yet, the company also learned that though the net response rate may be optimized, these new cardmembers may not be profitable. In other words, they may not be using the card or, worse, may simply cancel it. As a result, Amex decided that the goal of acquiring new cards should be based both on net responsiveness of prospects as well as on their overall profitability in the first year.

Solutions predicting response, attrition (cancel), and credit risk can be combined to provide an estimate or predictive value for an approximation of customer profitability. The word "approximation" or estimate is used because all fixed and variable costs and revenues are not factored into this definition of profitability. Adopting the approach of calculating profitability with all costs and revenues (fixed and variable) would consume an inordinate amount of time and be counterproductive to the goal. The goal in this exercise was not to arrive at an exact definition of customer profitability but rather at a more relative measure of profitability. In other words, the goal was to arrive at a profitability metric that could differentiate customers based on a relative measure. Even so, this required buy-in and consensus well beyond the analytics department. The departments of finance, marketing, operations, and systems were all involved as key stakeholders to ensure that this so-called profit metric would be recognized as a key measure when engaging with customers. At the end of this process, consensus was obtained among all key stakeholders regarding this profit metric. Recognizing that the profit measure was relative rather than absolute, employees then used it to assign customers to different segments, a very high group as the most profitable and a very low group as the least profitable.

Now that agreement was obtained on how to calculate this metric, the next part of the exercise was for analysts to create this measure at the level of the individual customer by using existing analytical tools. In this case, SAS (Statistical Analysis Software) was used as the analytical tool, and reports were built to see what this measure looked like when compared to other KBIs (key business indicators); an example of one type of report is shown in figure 2.2.

As the reports in figure 2.2 show, the profit metric does a good job of assessing retention performance, customer spending, and credit risk based on the rank-ordering of these KBIs against profit. The profit metric also performs adequately in assessing average time to make payment and average number of products bought, but it is not as useful as the other KBIs.

% of Customers Ranked by Customer Profitability	Index of Avg. Profitability per Decile	Index of Average Retention Rate per Decile	Index of Average Monthly Customer Spend per Decile
0–10%	5	4.5	4.2
10–20%	4.2	3.9	3.7
20–30%	3.2	2.8	2.6
30–40%	2.7	2.6	2.2
. . .			
90–100%	0.3	0.3	0.5

% of Customers Ranked by Customer Profitability	Index of Average Credit Risk per Decile	Index of Average Time to Make Payment	Index of Average # of other Products
0–10%	4.4	3	2.8
10–20%	4	2.6	2.4
20–30%	2.9	1.9	2
30–40%	2.7	1.6	1.5
. . .			
90–100%	0.5	0.75	0.7

Figure 2.2 Example of KBM Report and Customer Profitability

This kind of report provides valuable information to all the key stakeholders to help them determine whether the profit metric is a meaningful measure of profitability.

Once the profit metric has been defined and agreed upon by all the key stakeholders, models could then be built to predict profitability. At this point, analysts then had models to predict net response and profitability. Using these models, prospects could be evaluated based on their predicted ROI with Predicted ROI = (predicted profitability) / (predicted net response).

PREDICTED NET RESPONSE

This exercise took several years and finally produced the outcome and decision report shown here in figure 2.3.

As a result, the Amex marketing team could select prospects based not only on ROI but also on net response because one of their key performance measures was the number of cards in force. In effect, this decision

Predicted Net Response Deciles (1 being the best and 10 being the worst)

Predicted ROI (1 being the best and 10 being the worst)

	1	2	3	4	5	...	10
1							
2							
3							
4							
5							
....							
10							

Figure 2.3 Example of 10 × 10 Gains Chart Matrix

tool allowed marketers to demonstrate the impact of optimizing ROI, which would be at the expense of acquiring more cards.

This case study demonstrates how an organization developed its acquisition strategy based on data mining. The decision-making process in each stage was based on results and also on input regarding what the organization had to do to achieve the next level of improvement.

CHAPTER 3

DATA MINING IN THE NEW ECONOMY

IT IS IMPORTANT TO UNDERSTAND THAT THE SAME LEARNING FEEDBACK process used to generate strategy improvements can also be used in a more granular fashion by identifying specific tasks and tactics for improving future marketing efforts. Successful data mining practices include the application of ongoing learning to future activities.

Marketing environments must be learning environments based on prior learning. As shown in figure 2.1, the continuous feedback loop represents the underlying learning philosophy of any data mining exercise.

However, as with any business process, the key to data mining is having the right people in the right positions. Any investment in this sector should focus on people first and technology second.

Definitions of data mining abound. However, with the many experts and pundits offering their opinions on data mining, not much discussion has ensued concerning the roles and responsibilities of the people working in this field. Data mining has certainly impacted all positions in marketing. The position of data miner or data analyst is relatively new although traditional large direct mail companies, such as American Express and Reader's Digest, have also referred to these roles as modeling or statistical analysts.

Data miners succeed by serving as the conduit between the technology area and the business function area. In most cases, this functional area is marketing. However, other departments such as operations also utilize data miners, particularly in initiatives involving estimates of customer-level risk. To bridge the information and communication gaps between functional areas, data miners must often be generalists.

Yet, at the beginning of their career, data miners usually focus on technical expertise. Based on this technical expertise, organizations hire individuals who also have the acumen and adaptability to grow into the more generalist position. This is not easy because generalists and technicians appear to be very different, but successful organizations will hire persons with the right balance of both technical and general skills.

The seemingly contradictory sets of skills are best understood by describing the roles and responsibilities of data miners. For example, a challenge is presented to a group of key stakeholders in a business. Data miners must understand where data mining can have an impact. For example, one company may have retention rates of 10% for product X and 90% retention rate for products Y and Z. Clearly, the company has a much larger problem than the data mining one of identifying customers who are most at risk of cancelling product X. In fact, the priority would probably be to better understand the entire product offering rather than focusing on data mining efforts for better targeting of customers. Conversely, for example, company DEF has experienced tremendous growth in acquisition over the past 5 years, but the cost per new customer has significantly grown in the past 2 years. In this case, the data mining task of reducing cost per new customer through better targeting tools would apply.

These examples show that solid business generalist skills are required to identify the opportunity and challenge that can be impacted by data mining. With the challenge identified, data miners now need to draw on their technical skills and identify the appropriate data sources, technologies, and technical tools needed for a given solution. Once the solution is developed, the application of that solution requires data miners to serve in two roles simultaneously. For instance, they are business generalists during the design of an overall testing and tracking matrix to ensure that learning as well as business performance objectives are being optimized. At the same time, they serve in their technical role in applying this data to build this matrix.

Data scientists appear to represent a new business role and are gaining popularity as being critical to today's big data World. In effect, this role is identical to the data mining role. Yet, increased awareness of the data scientist role has placed more emphasis on the ability to bridge the gap between the technical side and the business side. Certainly, without the proper business insight into actioning solutions and the discipline to measure them from an ROI perspective, data miners will fail regardless of their technical expertise. A list of some data mining considerations relevant for developing a business-oriented solution follows below.

1. What is the business objective?
2. What is the desired solution and is it tactical or strategic?
3. What are the limitations of the data environment?
4. How will the appropriate data environment be constructed?
5. What are the business expectations?
6. What are the tools and technologies to be deployed?
7. How will the solution be applied?
8. How will results be tested and tracked?

We are all, to some extent, experiencing the drastic transformations underway in our current economy. Spurred by the subprime mortgage debacle, the world's economy seems to have folded like a house of cards, and significant job losses have followed throughout the world. In certain sectors, such as manufacturing, many of these jobs are gone forever. This major economic transformation is not new as economic historians can cite numerous times when the economy underwent major changes. In addition to this economic transformation and the requirement for corporations to become more knowledge-based, we are also witnessing social transformations. In particular, the focus is now on the environment and how we can be responsible stewards of its resources. For direct marketers, the challenge is to execute marketing program activities with significantly fewer resources. For example, untargeted direct mail campaigns are now an extinct business practice.

The new economy has also resulted in reduced corporate resources for doing a given task. Fewer people are available to do the same task— having to do more with less is the prevailing situation.

These economic and societal changes are resulting in demands for skills that transform data into meaningful business intelligence. Clearly, jobs related to data mining and analytics represent areas of growth in the new economy.

ROI: A Critical Output from Data Mining and Analytics

Even in today's economic downturn, work for data mining professionals seems not to have decreased and often has even increased. This is not surprising since a major demand of the new economy is accountability. In marketing, accountability is about measurement—the marketing mantra here is "What gets measured gets done." Establishing an ROI on every marketing initiative is now common for most marketers, and data miners and analysts are vital members of the marketing team. Data miners' ability to understand data and the marketing business needs allows organizations to create ROI templates for a variety of marketing initiatives.

DATA MINING SOLUTIONS TO OPTIMIZE ROI

In addition to accountability through measurement, the new economy also demands solutions that actually maximize ROI, as opposed to maximizing revenue, which used to be the traditional marketing objective for most of the twentieth century. The data mining industry also traditionally focused on developing predictive models to maximize revenue given a certain cost (e.g., response rate). For many organizations, data mining is still a daunting and formidable challenge rather than a common business practice. How can organizations change this? The key is to identify quick wins that demonstrate significant positive benefit within a very short time frame.

CUSTOMER VALUE AS A QUICK WIN

Let's look at some examples of quick wins. The first and most common quick win is for an organization to identify its best customers. Using the 80/20 rule, we can determine how company revenues are distributed among its customers. Using purchase revenue (which for most organizations is more than adequate in identifying customer value), we can create a measure of a customer's value by summing up each customer's purchase revenue over the past 12 months. Customers are then sorted into deciles (10 customer groups of equal size), with decile 1 being the group with the highest value and decile 10 the one with the lowest value. Figure 3.1 shows what a schematic of this would look like.

The production of this type of report (often referred to as a gains table or decile report) reveals that 60% of the company's revenue is delivered by 20% (deciles 1 and 2) of customers (not quite the 80/20 rule mentioned above). Customers in the top two deciles would represent our highest value customers. Medium-value customers would be represented in deciles 3 and 4 and account for 25% of all revenue. Low-value customers

Customer Decile	# of Customers	% of all revenue captured in decile	Segment Group
1	50,000	35%	High-Value Customers
2	50,000	25%	
3	50,000	15%	Medium-Value Customers
4	50,000	10%	
5	50,000	5%	
6	50,000	3%	
7	50,000	3%	Low-Value Customers
8	50,000	2%	
9	50,000	1%	
10	50,000	1%	

Figure 3.1 Value Segmentation Report

are found in the remainder of the file (deciles 5–10) and account for only 15% of all revenue. This information is critical to allow organizations to effectively deploy customer management strategies and tactics. The basic ability to prioritize marketing activities against different customer groups based on customer value will yield significant performance returns. For example, a greater proportion of resources should be devoted to customer segments that deliver the highest returns.

CHANGE AS A QUICK WIN

Another quick win concerns changes in customer behavior. Marketers historically have been interested in identifying changes in customer behavior and, in particular, in changes pertaining to lifecycle. The key is to identify these lifecycle changes based on what is known about customer behavior and demographics. Knowing that a given customer is a university student and about to graduate would represent key information to a marketer for designing the appropriate programs that might appeal to this person. For example, if a customer has just turned 65 and significantly changed certain expenditures, such as spending less on gas and more on travel, we could infer that this customer may have become a retiree. In other examples, we might see a married person significantly increasing his or her purchases on categories related to furniture. This might suggest that this person is perhaps buying a house. This same married customer might purchase more baby clothing, and we can infer that this person has now become a parent. In all these examples, data is required to make such inferences. Inferences can be made with any data, but the type of data available dictates how confident we are in our inferences and how we use these insights. For example, a change in an investment portfolio with a customer aged 30 going from a risky portfolio to a less risky one might suggest a number of events; the customer might be getting married, starting a family, or simply just changing his or her investment strategy. In any case, without even recognizing the precise reason for the change, the marketer would develop strategies and tactics to address the fact that there has been some change.

Engagement as a Quick Win

Change and value are both good quick wins from a data mining perspective, yet another quick win relates to customer engagement. Although engagement can be similar to value, this is not always the case. A high-value customer of a bank may have a large loan and large mortgage but make all payments through automatic withdrawal. If this is the way the customer interacts with the bank, the customer's engagement level is very low. Other information must be considered to obtain a true reflection of customer engagement. With the growth of digital marketing, the extent of customer engagement becomes much broader as we can now look at customer-initiated activities, such as e-mail communications, web site browsing activity, and inbound telephone calls and other means of customer-initiated communication with the company. This communication activity together with purchase activity and response activity across all channels can be integrated to create a more complete measure of customer engagement.

In today's new economy, data mining and analytics are becoming a common business discipline. Since the discipline is still new, many people consider data mining and analytics simply more advanced statistical techniques. This misperception has made the discipline seem more complex even though simple solutions using the data in a pragmatic way often deliver great business results.

When assessing and prioritizing projects, analysts and marketers generally combine projects into what is called "low-hanging fruit" or exercises that can yield great benefit quickly. These projects are usually those on which no data mining has been done in the past, and a significant benefit can be achieved in a very short time and often with minimal resources.

The "high-hanging fruit" can also achieve significant benefits but require more resources and commitment. In most cases, practitioners are working in projects on which some data mining has already been done, and they work to increase the benefits compared to what the existing data mining solution provided.

Success depends on people. Nowhere is this more applicable than in having the right team for a given project. Projects will be successful if we take a multidisciplinary approach combining expertise in mathematics, statistics, and databases with the domain knowledge of the business in which the data mining solution will be applied.

Organizing and managing this information is another critical element to success. The benefit of using high-powered mathematical solutions depends on the quality of the inputs (data and information) used.

In data mining, practitioners must always be cognizant of the cyclical nature of projects. That is, the practitioner must not only work to maximize results but also to obtain more learning once a given solution is applied in a specific marketing initiative.

No project will be successful without a business champion. This person is not the data mining practitioner but typically a key individual in marketing who takes ownership of all the results even those where data mining tools are applied. This person's vested interest in the success of data mining makes him or her an "evangelist" for these solutions throughout the organization.

CHAPTER 4

USING DATA MINING FOR CRM EVALUATION

AS STATED IN EARLIER DISCUSSIONS, DATA MINING IS NOT JUST ABOUT building tactical solutions for specific issues, such as the creation of a retention model for new Telco customers. Data mining can also be used to develop overall strategies, such as a broad segmentation strategy. At an even more general level, data mining can be used to determine whether the deployment of a CRM program makes sense. Measurement and evaluation play a role in the following areas:

- Evaluating CRM in general
- Developing an overall segmentation strategy

EVALUATING CRM IN GENERAL

In this scenario, the use of a pilot project or proof of concept can provide the necessary information very quickly and with minimal investment.

At this basic level, the goal is to determine whether or not a CRM strategy will yield a positive ROI. Results from a small pilot test can be extrapolated and eventually rolled out as a CRM strategy. The following is an example of this process.

Company ABC is unhappy with overall customer retention and is considering various options, such as the viability of a CRM retention program. In the pilot project test, results show that a retention program can increase retention rates by 2%. This then becomes the key metric in creating the spreadsheet (figure 4.1). Without a retention program, the

	Year 1	Year 2	Year 3	Year 4	Year 5
No retention program					
Number of Customers	10,000	8,500	7,225	6,141	5,220
Average Profit/ Customer/Year	$300	$300	$300	$300	$300
Net Profit/Year	$3,000,000	$2,550,000	$2,167,500	$1,842,375	$1,566,019
Year over Year Retention Rate	85.00%	85.00%	85.00%	85.00%	85.00%
Retention Program					
Number of Customers	10,000	8,700	7,569	6,585	5,729
Cost of Retention Program/Customer	$5	$5	$5	$5	$5
Total Cost of Retention Program	$50,000	$43,500	$37,845	$32,925	$28,645
Average Profit/Customer/Year	$300	$300	$300	$300	$300
Total Profit	$3,000,000	$2,610,000	$2,270,700	$1,975,509	$1,718,693
Net Profit/Year	$2,950,000	$2,566,500	$2,232,855	$1,942,584	$1,690,048
Year over Year Retention Rate	87.00%	87.00%	87.00%	87.00%	87.00%
Cumulative Differences					
Cumulative Net Profit No Program	$3,000,000	$5,550,000	$7,717,500	$9,559,875	$11,125,894
Cumulative Net Profit With Program	$2,950,000	$5,516,500	$7,749,355	$9,691,939	$11,381,987
Cumulative Incremental Profit with Program	($50,000)	($33,500)	$31,855	$132,064	$256,093
Cumulative Incremental Cost of Retention Program	$50,000	$93,500	$131,345	$164,270	$192,915
Cum ROI	–100%	–36%	24%	80%	133%

Figure 4.1 Evaluation of Retention Programs by 5-year ROI

retention rate is 85%, but with a retention program the rate increases to 87%. The other assumption of profit per customer of $300 is held constant in both cases.

In building this spreadsheet, one might consider a short time frame, such as one year, in which to measure the impact of a retention program. However, in evaluating an overall strategy, one needs to view the impact over a longer term, such as five years. For more tactical projects, such as comparing the impact of two communication offers in a campaign, a time frame of one year is more than sufficient. In the example of the CRM retention strategy, a time horizon of five years is used since this represents an attempt to keep customers for the long term and not just for one year.

From the spreadsheet, which uses a base of 10,000 customers, one can see that the company may actually lose money in the first two years on a cumulative basis, but it will begin to make money in the third year through the cumulative impact of the increased retention rate from 85% to 87%. As one would expect, all kinds of what-if analyses can be done by varying the cost of the retention program. In addition, a range of improved retention rates around 2% could be considered in order to determine the overall profit and ROI. Figure 4.2 shows a sensitivity analysis on a range of improved retention rates around 2%.

Although this example may seem simplistic to produce, it is important to remember that simplicity is essential in conducting these types of analyses. While producing these spreadsheets, metrics or measures need to be determined that are affected by the program and that are truly incremental. As seen from the above spreadsheet, what-if sensitivity analyses can be very useful in revealing the range of results that could be achieved based on different scenarios. In this example, the results can range from $32K at 1% improved retention rate to $954K at 5% improved retention rate. Given this large range of results, it is important that the analyst design a pilot project that produces results

Improved Retention Rate	Cumulative Profit	ROI
1%	$32,966	17%
2%	$256,093	133%
5%	$954,651	466%

Figure 4.2 5-Year CRM ROI at Different Improved Retention Rates

Improved Retention Rate	Cumulative Profit	ROI
1.9%	$233,565	121%
2%	$256,093	133%
2.1%	$278,669	144%

Figure 4.3 5-Year CRM ROI Bounded by 95% Confidence Level Around 2%

with a minimum error range. For instance, if the improved retention results are 2% and the range of accuracy is at 95%, then the confidence level is 1.9% < = 2% < = 2.1%. It can then be inferred that 95 times out of 100, the expected improved retention rate will be between 1.9% and 2.1% as the program is rolled out. When reconsidering the above sensitivity sheet within this error range, more precise results can be provided to senior management (figure 4.3). The profit in this case ranges from $233 million to $278 million, a range with far more precision than the original one of $32 million to $954 million.

Although this example dealt with only one metric (retention rate) that was incremental, in many cases the introduction of a CRM retention strategy can impact more than one metric. In the same example, overall customer spending is also increased by $2 each year.

In this scenario, by impacting retention (2%) and spending ($2 per year), the cumulative profit by the fifth year has been increased from $256 million to $466 million.

A meager increase of $2 per year has almost caused profit to double, which simply reinforces the notion that small numbers can provide huge payback when the results are viewed in a long-term horizon (five years).

What if a third element is added to the picture? Suppose the company decided to develop a retention model that allows it to target high-risk defectors in a very cost-effective manner, assuming that the model will still achieve the same save rate, which is increasing the retention rate by 2%. The assumption of a constant save rate (2%) between untargeted names and targeted names is a large one. This assumption would need to be tested through back-end analysis of save rates between targeted and untargeted names. For purposes of this exercise, assume that the save rate is constant between targeted and untargeted names. The measure that is impacted by the model (incremental) is the cost per customer. In this case, it turns out that the overall cost per customer can be reduced from $5 to $3.

With a targeted program, the cumulative incremental profit has now increased from \$466 million to \$543 million. If the cost of targeting is less than \$80 million, then adopting a more targeted approach to retaining customers is worth the cost.

This page intentionally left blank

CHAPTER 5

THE DATA MINING PROCESS: PROBLEM IDENTIFICATION

WITH CRM BECOMING MORE OF A BUSINESS PRACTICE FOR MOST organizations today, data mining is often viewed as the analytical technology component required for achieving a given solution. Does this describe the true nature of data mining? Many people consider data mining as a technological component and think that purchasing the right software and hardware is the key to effective data mining. Other schools of thought regard the use of statistics and/or computer programming together with machine-learning algorithms as the essential component of data mining. Yet, these preconceived notions miss the mark because they focus on specific components in the overall data mining process. Data mining is a step-by-step process requiring humans to interact with technology in order to produce the best business solution for a given process. So what is this data mining process?

In this book, we divide the data mining process into four major steps or stages, as outlined in figure 5.1:

1. Identification of the business problem or challenge
2. Creation of the analytical file
3. Application of the appropriate tools and technology
4. Implementation and tracking

Experienced practitioners realize that problem identification is probably the most important stage in the entire process. Ironically, despite the focus

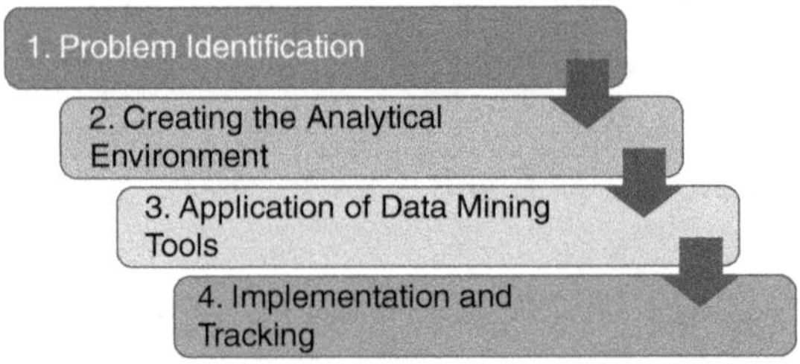

Figure 5.1 The Four Stages of Data Mining

on the many data mining technologies available in the marketplace, at this important stage of the process the human element is most important.

What matters is the expertise of critically assessing a given business situation quantitatively and qualitatively and to determine its importance given the overall business strategy. For example, an organization's overall sales had significantly decreased within the past year, and 75% of the company's overall sales resulted from 20% of its best customers. The analysts and marketers determined that customer attrition had significantly increased in this past year.

With this information, it was decided that an attrition model to identify high-risk attritors would allow marketers to allocate more resources to this vulnerable group in the hope of retaining these customers. Yet, this thinking was flawed for two reasons. Analysis should have been conducted on the high-value risk group to determine the extent of the attrition problem in this segment. Based on the preliminary information on this specific business case, it is likely that the attrition problem is highly prevalent in this segment. The other thing to consider before developing a model is to understand whether CRM or data mining has any relevance in this situation. For instance, if the reason for losing sales is that the competition has created new products or services with price points and benefits superior to those of the company in question, then even superior CRM practitioners with their data mining practices cannot stop this sales erosion. Clearly, some up-front market research is the preferred route at the present time in order to better understand why this situation is occurring.

Another important consideration in identifying a problem is to understand the current data environment and its implications for resolving a particular business problem. For example, a company may want to launch a new product and develop a cross-sell model to target customers who are most likely to purchase this product. In building a model, analysts need to understand that there is no prior information or history on this specific product with which to develop a model. The analysts might then look for other similar products with previous history that could be used to develop a broad profile. If this is not the case, analysts would have to think of ways to identify customers who might be early adopters (i.e., the pioneer purchasers of new products).

Often enough, business users have an idea of what they want to derive from data mining and have a solution in mind but do not understand the long-term ramifications. For instance, a credit card company may want to acquire the most cost-effective prospects. From the business users' perspective, this may simply require developing a response model. From the perspective of experienced data miners, this request can seem quite vague. Experienced data miners can quickly point out a number of specific questions about the particular solution to be developed. For instance, should a simple gross response model be developed? This type of solution will yield cost-effective cards in the short term, but in the long term the creditworthiness of these new cards will be severely eroded. Although the marketing cost per new card may be dramatically lowered, this benefit may be outweighed by increased operations costs of processing these bad new cards. The solution to this dilemma is to build a solution that optimizes both gross response rate (marketing cost per card) and approval rate (operations cost per new card). Yet, a further dilemma may be a high level of attrition in the first 6 months after acquiring the card. In this case, the business must focus on achieving the following three objectives:

1. Maximize gross response rate with the ultimate objective of optimizing marketing costs
2. Maximize approval rate with the ultimate objective of optimizing operations cost
3. Reduce attrition rate with the ultimate objective of increasing the overall level of profitability in the first year

A similar example can also be seen in the telecommunications sector where acquisition efforts are conducted through outbound telemarketing. Despite the efforts to become more efficient through response models, the organization still encounters the negative dynamics of excessive customer defection in the first three months. Experienced data miners would recommend solutions that provide a balance of acquiring new prospects in a cost-effective manner with optimizing the retention rates of these prospects once they become customers.

As these examples show, some creativity in thinking needs to be employed to ensure that the right problem and challenge are identified in the given business circumstances. Collaboration between the marketing or business area and the data mining area is essential for optimizing the creative thinking from both functional areas

How does this collaboration occur? A meeting of the data analysts and the marketers is often the start. Business users outline the problem and their vision of the solution, a vision that is often not inappropriate for the challenge at hand. However, this vision may be inappropriate or not allow for other options that may be superior to the one selected initially. This is where data miners need to probe and ask more questions to obtain as much as information regarding the business issue as possible. For instance, they should ask questions such as the following:

1. What is the overall marketing strategy and business objective?
2. What is the specific objective of this initiative?
3. What are the marketing plan and budgets and forecasts?
4. What historical information and reports do we have?
5. What is the data environment?
6. What are the product dynamics and competitive/economic forces that will impact the overall results?

What Is the Overall Marketing Strategy and Business Objective?

The overall marketing strategy essentially is an initial guide indicating the extent and complexity of data mining that will be required in a given organization. For instance, the strategy may be one of tremendous growth within a certain sector. The company may tackle this challenge on two fronts. First, it may already have existing products or services that meet the needs and expectations of customers in this sector. Yet, its overall presence within this business sector may be limited, and investment in marketing efforts to increase the company's presence and awareness of its existing products and services would be a priority. Second, the company would want to investigate new products and services to further enhance its overall offerings to customers in this sector.

A good example of this would be credit card companies offering insurance services to their customers. Initially, the card may have offered travel insurance or protection in case the card was lost. With this customer base, the company has an opportunity to offer other types of insurance either by itself or with business partners. Such companies would invest in programs to increase awareness among travel insurance customers and credit card registry (lost card protection) insurance. At the same time, funds would be invested in extensive market research to identify appropriate insurance products and services relevant for credit card customers. Funds would also be invested in test-marketing the findings and develop an overall launch program. All these efforts are undertaken to meet the strategic imperative of significantly growing the insurance business among this customer base in order to become a significant player in this sector. From a data mining standpoint, the key phrase to focus on is significant growth because this implies that costs are not a major priority. With this objective, companies are willing to sacrifice inefficiencies in order to increase their overall market presence. The use of data mining in this scenario is not a priority and would be employed sparingly.

However, this will likely change once the initial launch has been completed. As acquisition and growth become an ongoing activity, the need to do it in a more cost-effective manner will necessitate the use of data mining. The key to when data mining becomes most applicable is when a given marketing activity is considered ongoing but its costs are increasing at a faster rate than revenues.

Another critical strategy might be a company's focus on retaining customers. Data miners must understand that if the focus is on retention, then cost-effectiveness may or may not play a role. For instance, a given company may have a highly selected group of high-wealth customers. These customers number about 5,000 and represent 5% of all the customers, but they account for 50% of all revenues. The company may find that attrition rates have increased by 50%. In this case, data miners could challenge whether the identification of high-risk high-wealth defectors through data mining will provide real business value. In other words, identifying who might be likely to defect is not as important as understanding why these customers defect. That is, business owners must understand the key motivations/attitudes and the competitive forces causing large problems with retention. As these examples show, probing to better understand the strategy and its objectives provides an initial guide as to the overall impact of any data mining solution.

It is often easy to use data mining as a quick hit solution with an obvious immediate net benefit to the company. The demonstration of these results to an organization's executive team represents an easy avenue for managers to show their value to the organization. However, these results, despite their net immediate benefit to the company, may actually disguise more long-term business problems that will severely impact the balance sheet in a few years. For example, analysts may build retention models that allow marketers to target high-risk defectors but hide the fact that overall interest in the company's products and services is declining.

Aside from the retention example described above, another good example is the introduction of a new product. For example, using some basic data mining tools, companies can achieve an overall net benefit to the marketing program. But even with data mining, it may still be true that in some scenarios a company's profitability has been seriously eroding over the past few years. New competitors have emerged in the marketplace, and severe price competitiveness has resulted. The company has maintained its pricing strategy and ultimately its overall contribution margin on a per-unit basis. However, the fixed-cost proportion of overall expenses relies on a certain amount of sales volume to meet these expenses. With price competition, the volumes have deteriorated significantly thereby increasing the fixed-cost component as a percentage of revenue. At this point the company should focus on repositioning the product. Rather than get into the specifics of what this entails since that is neither the purpose of this book nor my expertise, it is enough to say that data mining at this point in time is clearly imprudent and would use up resources that should be devoted to other areas of the business.

What Is the Specific Objective of This Initiative?

In many of the initiatives, the solution is tactical as the marketers or business persons already have a preconceived notion of what they want to do. For instance, the development of a cross-sell model represents the type of solution to cost-effectively promote other types of products to existing customers. In another situation, the objective may be to lower the costs of acquisition efforts by building response models. On the surface, the decision to build these tools may make perfect business sense; however, management must determine whether these short-term solutions are consistent with the overall marketing strategy. If the strategy is to provide significant growth, the use of data mining would not be considered a key component in the initial phases of this strategy. Awareness rather than cost-effectiveness is the prime objective of such a strategy. Once this is achieved, maintaining awareness and continuing growth become more challenging due to such forces as market saturation and increased competition. At this point, data mining may indeed be more relevant as marketers strive to do this in a more cost-effective manner. If the specific objective is consistent with the overall strategy, analysts have to think through the preconceived solution. For instance, they could use a cross-sell response model to promote a specific product to the right group of customers. But the question is whether this is the right product. Despite having a high predisposition toward responding to an offer for this product, customers may have a higher predisposition toward another product type. Product affinity models provide a solution whereby products can be ranked in relative terms of their purchase affinity for a given customer.

In some cases, identifying the business problem or challenge is considered to be predetermined since this occurs before the actual mining of any data takes place. With this approach, data mining is seen as a purely tactical approach that develops tactical solutions for specific problems and challenges. Organizations following this approach assume that data mining either plays a very limited role or none at all in the development of strategy. It is my belief that this type of approach is less than optimal. Organizations that use data mining for both strategic as well as tactical solutions will achieve superior solutions than those using data mining purely for tactical purposes. Including problem identification as part of the data mining process acknowledges that data mining can be used to derive insights for the development of new strategies. Let's take a look at a few more examples.

For example, a company is experiencing significantly eroding sales from its outbound telemarketing acquisition efforts. Their current acquisition strategy has not changed over the past couple of years. Based on this cursory information, it would appear that list fatigue has set in, and analysts need to develop tools to produce more targeted lists. Accordingly, marketers have instructed the data mining department to build a response model to help optimize sales.

But the lack of collaboration between the data mining department and the marketing department ultimately results in a solution that does not take into account the big picture. The acquisition model is built with the intention of maximizing response. As a result, response is maximized, but the sales numbers are still suboptimal. Further investigation reveals that the cancellation rate in the first three months among new customers has increased. Both marketing and data mining departments made assumptions here that the optimization of acquisition response would lead to optimized sales. If a more up-front and collaborative investigation had been conducted, it would have been found that a tool had to be developed that optimizes response and also the likelihood of retaining customers for more than three months.

Another good example of being reactive (tactical) rather than proactive (strategic) is when marketers claim a predictive model cannot be developed until a marketing database or data mart is set up. This decision delays the development of tools that can achieve tremendous cost savings for current campaigns. Data mining can yield solutions from raw legacy systems and databases in a very short time frame (in as little as four to six weeks).

Another classic example is the case of "we don't have the data." Letting data miners examine the data allows them to use their expertise and experience to provide an approach that leads to higher yielding results. For instance, acquisition programs provide the most varied approaches for targeting names. In some cases, analysts may develop individual-level models from companies that compile lists of names containing demographic and attitudinal information. Information pertaining to individuals is collected when consumers fill out a lengthy survey in return for coupons that can be redeemed to purchase more products at a discount. Other individual-level sources for acquisition contain information relating to consumers" magazine purchase habits. Here, the information concerning magazine purchase behavior is compiled into a magazine category. The actual fields of data then reveal the recency, frequency, and value of purchases in an overall magazine category. In addition to the magazine purchase information, Stats Can information in the shape of PSYTE demographic clusters (60 clusters nationally) can be appended to each individual record.

In other cases, analysts may develop geographic-level models if only name and address information is available. These acquisition models utilize Statistics Canada demographic information, which is aggregated to a particular geographic level. For census data containing information such as ethnicity, occupation, education, and a range of other population demographics, the census level information is aggregated to approximately 400–500 households. The disadvantage of this information is that it is collected only once every five years. Over time data becomes stale, which is a problem. Meanwhile, the postal walk data collected from personal tax returns contains information about tax filers related to their income and wealth. This information is collected annually and thus is definitely superior to the somewhat stale census data. However, its limitation is its lack of granularity because the data is aggregated at the level of the postal walk, which represents approximately 800 households and is thus inferior to the more granular enumeration area EA census tract of 400 households. Furthermore, the breadth of data is limited as information is primarily related to income and wealth and does not cover as broad a range of information as census information. The use of a conversion model table, though, whether at the census or postal walk level, is critical to mapping the appropriate demographic data back to the postal code level. The table in figure 5.2 shows an example of some demographic Stats Can data mapped back to census and to postal code.

The postal code information represents the critical link in appending this data back to either prospect or customer files.

Another limitation when dealing with Stats Can census data is the government's most recent decision to make the census totally voluntary. This means that when analytics exercises compare information over time or between different time periods, the reliability of this information is questionable. However, if the data is used for rank-ordering purposes, then reliability is less of an issue in producing an effective data mining solution.

Although the above issue is specific to Canada, the approach and challenges in dealing with external geographic area data are similar in all countries. Challenges in understanding the granularity of the data or the

Census	**Postal Code**	**Median Income**	**Avg. Age**	**Avg. Household Size**	**% of population with university education**
100001002	M5A2J1	$42,000.00	40	2	10%
100001002	M5A3J3	$42,000.00	40	2	10%
100001002	M5A4J1	$42,000.00	40	2	10%
100005004	H4B2E5	$50,000.00	35	1	85%
100005004	H4B3E1	$50,000.00	35	1	85%
100006008	L1W3K6	$37,000.00	43	3	5%
100006008	L1W2L5	$37,000.00	43	3	5%
100006008	L1W3K1	$37,000.00	43	3	5%

Figure 5.2 Example of Geographic Table (Census and Postal Code) and Demographic Variables

geographic level of data capture are consistent across all countries. The question is whether this data can be appended back to customer files?

These geographic-level models using the Canadian example, whether derived at the census level or the less granular but more frequently updated postal walk level, will never be as powerful as models developed from individual-level data, but they can still yield very powerful results especially when applied to list universes that are very large (i.e., more than 1 million names). These models in many cases can be more broadly applied since they are developed at the level of a geographic area. Meanwhile, individual-level models can only be applied to those lists from on the basis of which the model was originally developed. Moreover, these lists may have inherent skews or biasness in the data. For example, models based on responders to a coupon survey are an example of this kind of biasness. Comparing the responder pool to the Canadian population at large, analysts may learn that the typical coupon responder pool tends to skew toward those with lower income and females. Companies using this biased data in order to build acquisition models would need to understand this dynamic before even considering building any models. In all likelihood, the overall list representing this source of data would be tested regarding its potential to achieve an acceptable response rate. If this threshold was met, then the list would be explored further in terms of using data mining techniques such as acquisition models to optimize the overall results.

Another example of creating targeting tools for acquisition purposes is the use of customer penetration indexes that are based on the principle of "fish where the fish are." Using penetration indices as targeting tools is both pragmatic and easy to understand and, more important, easy to do. No deep understanding of statistics is required for deploying these tools.

In some cases, very limited information is available to help resolve business problems. But that does not mean that data mining techniques cannot be employed. For example, let's assume a retail company would like to initiate a CRM discipline in its acquisition programs. First of all, the company would like to use data and information to improve its overall effectiveness. The only information or learning available is from the company's research.

This retail company collects no information on its customers. Market research has indicated that the key drivers of purchase behavior are high income, female immigrants.

Here, data miners and marketers can think of creative ways to capitalize on learning that currently exists rather than expend resources in a campaign and to also be forced to wait another three to six months before obtaining the results of this campaign. The table in figure 5.3 outlines

	Income	% Female	% Landed Immigrant
Average Postal Code	$40,000	52%	5%
M5C 1J2	$50,000	60%	10%
Index	1.25	1.15	2

Figure 5.3 Creating a Postal Code (Geographic Area) Level Index

how the creative use of Stats Can information as a proxy for research learning might provide one targeting approach. Each of the key pieces of market research learning represents available information that has been compiled at the enumeration area by Statistics Canada. Using a conversion table (enumeration area to postal code), analysts can then create fields of this key information at a postal code level. From this table, analysts can create an overall index based on this information. The table in figure 5.3 depicts how this overall index might be calculated.

Assuming that each key field or piece of information is equally important in the overall index, then the index for M5A1J2 is

$$(.33 \times 1.25) + (.33 \times 1.15) + (.33 \times 2) = 1.45.$$

Before adopting this index approach, the appropriate tests and controls would be put in place to evaluate the current approach and to develop new learning. This new learning could also be utilized to develop acquisition models if the appropriate random samples are created within the test matrix. For the future, analysts could determine which approach is superior (acquisition model or penetration index). The key point here is that current learning can be acted on immediately rather than having to wait for the outcome and results of the next campaign.

Leveraging both the marketing expertise and the data mining expertise in identifying problems helps to ensure that the organization is more clearly focused on the higher priorities with more significant impact on the business. The marketers' experience in terms of prior campaign results and strategies complements the data miners' experience and expertise in terms of understanding the data environment and what works best for a given business situation. This combination of skills represents a competitive advantage in identifying business problems or challenges that can potentially yield greater returns to the business.

However, problems or challenges are often entirely open-ended, and data mining then becomes an exercise in discovering knowledge or data that provides direction and strategies for activities that should be undertaken over the course of the next year or two. This process involves a staged process, and the key stages are the following:

- Data audit
- Business results gathering and key stakeholder interviews
- KBM and post-campaign results
- Segmentation and profiling

We will discuss these stages in more detail in later chapters.

CHAPTER 6

The Data Mining Process: Creation of the Analytical File

Once the business challenge or problem is identified, analysts have to understand the data and information requirements for conducting the necessary analytics. Analysts do not need to perform a rigorous data needs analysis as would be required for building or designing a database; they are only concerned with what is already there and not with what should be there. When a file and some of its contents have been identified as potentially relevant for an analysis, analysts in most cases request the entire file as one data component in the project.

Other files would also be requested depending on their relevance to the project and whether customer-related information can be extracted. This is when the data audit process begins; it is a rigorous data exercise with the following objectives in the context of optimizing the data environment for analytics:

- Quality of data
- Data integrity
- Usefulness of data for data analysis

When the source files have been determined, analysts then must understand the quality of the data. In other words, are there certain fields that have a large proportion of missing information?

In the example shown in figure 6.1, 43% of the customer base has no start date as a customer. A number of techniques can be used to handle these missing values. For instance, using the average or median value of

Tenure	# of Customers	% of Customers
1998	49,000	14%
1999	49,000	14%
2000	59,500	17%
2001	42,000	12%
Missing	150,500	43%
Total	350,000	100%

Figure 6.1 Frequency Distribution of Customer Tenure

Product Category Code	# of Customers	% of Customers
ABC	103,810	30%
DEF	118,650	34%
GHI	74,165	21%
999	49,875	14%
Total	350,000	99%

Figure 6.2 Frequency Distribution of Product Category Code

the values not missing to impute an overall value is a popular way of dealing with missing values. A more robust but much more time-consuming way is to build a model or algorithm that predicts the value of a variable based on other fields or characteristics in the database. Another example of a poor data quality is when all records have only one outcome. For example, using gender as a potential variable in sports analytics involving major league baseball would be meaningless since the analytical file would consist of males only. We cannot draw any insight regarding gender as there are no females for comparative purposes.

Another problem concerns the integrity of the data pertaining to values in a field that don't make sense. In the example in figure 6.2, all product codes are comprised of letters and likely relate to a specific product category. The product category "999" suggests that some investigation is required to better understand what this relates to.

After the data quality and data integrity issues are ironed out, analysts must consider how certain fields should be summarized or grouped. This is particularly relevant for purchase history. For instance, to summarize the spending into yearly categories with one category for the customer's overall lifetime spending. At the same time, there may be hundreds of different product purchase codes. The challenge in using this data in a meaningful way is to group these product codes into broad categories so that the grouped information has enough data for any future statistical exercise in a given data mining project.

Once analysts have determined how to handle data quality issues and how to group or summarize the data, algorithms are then written to organize the data and information into one overall analytical file. This stage is often one where data miners or analysts best demonstrate their value to the organization by using their knowledge of the information environment to create meaningful variables or fields of information that will be most relevant for a given analysis. For instance, trend variables (growth and decline) as well as purchase variables related to time and type of purchase are derived from analysts' work and are not directly obtained from the source database files. In fact, the proportion of sourced (direct from the source files) to derived variables (created by analysts) is about 10% to 90%. Creating derived variables that are meaningful in the data mining exercise is one of the most important skills of data miners.

The ability to merge different files into one overall file is a very critical component of this exercise. This requires analysts to understand how to link files together through some preexisting match key or to develop the appropriate logic in creating this match key. The merging of different files involves a deep understanding of the various data relationships between files such as one-to-one, many-to-one, and many-to-many. Data can be looked at from two perspectives:

- Behavioral (how are persons or the record of interest behaving)
- Nonbehavioral, such as demographic data, external geographic area data, or activities that are impacting the person or record of interest.

Yet, the ultimate objective here is to determine which variables are going to be useful for the data mining exercise at hand.

DATA

Data mining practitioners agree that data mining success is based on the data. Although the use and knowledge of statistics is important, the lack of decent and reliable data will result in suboptimal solutions. Intimate knowledge of the data environment is the key to producing effective solutions. "Intimacy" may sound like an odd word when dealing with data, but it does express a certain level of closeness needed when conducting data mining exercises. This feeling of closeness implies that analysts must fully understand the reliability and integrity of each piece of data considered for a given data mining project.

The example of the education process can best serve to illustrate the importance of data. In the education system, reading and mathematics are considered essential requirements in any successful career. Yet, these two disciplines have basic technical fundamentals that must be mastered before students can be successful in either discipline.

In mathematics, students must master the fundamentals of addition, subtraction, multiplication, and division. In reading, they must master the alphabet before they can read a story. This process for mastering a particular discipline is much like the one for mastering analytics. In analytics, our ABCs are data. Analytics practitioners must have a conceptual understanding of data and a hands-on approach to data. With this approach analysts gain a much deeper understanding of the data and are better able to handle all the detailed nuances that are inevitable in the data world.

Conducting a Data Audit on the Source Data

In any environment, whether digital or offline, analysts must have a process for better understanding data. How is this done? The first step is the data audit, which we will discuss now.

Once the data sources have been identified for a given project, they must be mapped out to the source data in the source files in their IT environment. The appropriate source data is then extracted, and a data audit is conducted on this data. The extent or detail of this data audit depends on whether the project is in development mode or simply reproducing these reports on an ongoing basis.

If this is the initial creation of these reports, the source data and source files are likely to be unknown to the analysts. In this case, the initial development of these reports requires that the more exhaustive data audit be performed in order to explore each field or variable from all the source data. These reports provide insights on the following:

- Extent of missing values
- Range of values

- Average, minimum value, maximum value
- Number of unique values

The results from the data audit indicate the usefulness of a given variable in the overall set of measurement reports. For example, if age is a desired field in this measurement template but the data audit uncovers that 75% of the age fields are missing values, then the usefulness of age in reporting is severely limited by the missing values. However, analysts could create a binary variable from age based on whether the value is missing or not missing. Analysts have often found that this type of information can yield valuable learning in any measurement or analytical exercise.

Regardless of whether the data represents new information unfamiliar to the analysts or represents a familiar data source, a process such as a data audit is needed for a better understanding of the data at a detailed level.

The first task in the data audit process is to obtain an initial glimpse of the data by observing a random set of 10 records and all their accompanying fields from a given file, which is referred to as a data dump. This simple task accomplishes two objectives. First, it determines that the data has been loaded correctly into the analytics application used. Second, analysts get an initial sense of the data environment. In other words, is this a very simple data environment or one that is quite exhaustive?

In this example shown in figure 6.3, something has gone awry as the third record reveals values that do not make sense for all the fields that are to the right of the birth date. Further investigation reveals that missing values were not handled properly when loading the data into the company's analytical environment. Fixing this problem reveals what is shown in figure 6.4.

Here the number of files and the number of fields in each file together with examples of the values of these fields allow analysts to gain a very deep, detailed understanding of the data environment.

Account Number	Postal Code	Birth Date	Start Date	Behavior Score	Income	# in House
123456	M5A 3S6	Jul-49	Mar-91	500	30000	6
345321	H3A 2B5	Aug-54	Apr-92	550	42500	1
543235	T5A 2S7	Jun-83	600	35,500	3	543210
etc.						

Figure 6.3 Sample of 3 Customer Records and Fields

Account Number	Postal Code	Birth Date	Start Date	Behavior Score	Income	# in House
123456	M5A 3S6	Jul-49	Mar-91	500	30000	6
345321	H3A 2B5	Aug-54	Apr-92	550	42500	1
543235	T5A 2S7		Jun-83	600	3500	3
etc.						

Figure 6.4 Sample of 3 Customer Records and Fields, Adjusted

Variable	# of Records	Data Field Format	# of Unique Values	# of Missing Values
Income	100,000	Numeric	50,000	2,000
Customer Type	100,000	Character	4	10,000
Gender	100,000	Character	3	50,000
Household Size	100,000	Numeric	7	90,000
Product Type	100,000	Character	3,000	5,000
Customer Name	100,000	Character	100,000	0
Postal Code	100,000	Character	50,000	0

Figure 6.5 Data Diagnostics Report

Then, further data diagnostics are conducted so analysts can determine the number of unique values and the number of missing values for each field on each file they have received. An example of this kind of report is shown in figure 6.5.

This example shows that household size may not be a useful variable for an analytics exercise because 90% of the records have missing values. At the same time, product type, which has few missing values, contains over 3,000 different values. This would be no problem if the variable was continuous, such as income or spending. But in this case, the data format is character, namely, a product SKU, which is the typical most granular product level detail. Analysts must treat each outcome as a yes/no variable (all 3,000 outcomes become yes/no variables) where a yes is coded as 1 (presence of product SKU) and no is coded as 0 (no product SKU). What does this mean? If there are 100,000 records with 3,000 product SKUs, a rough average would indicate that 33.3 records out of 100,000 would have a specific SKU for a penetration value of .03%. The use of binary variables in this scenario would be meaningless in any future data mining exercise because there are too few 1s and too many 0s. Further diagnostics, such as frequency distributions, allow analysts to better understand the distribution of values in a given field. Some examples of such reports are shown in figure 6.6.

In the report in this example, it would be difficult to establish personal communication with these people (50% of customers) since their address is not known.

In the example in figure 6.7, gender might be a difficult variable to use analytically because 50% of the variable values are missing. Yet, this would not be the case for income where only 2% of the values are missing.

Another example based on tenure (creation date) is shown in figure 6.8. Here, customer counts increased significantly between 2005 and 2006, which indicates that acquisition in this period increased significantly. This would warrant investigation as to why this occurred.

With the information from these data audit reports, analysts can then determine how to create the analytical file that represents key information

Region	# of Customers	% of Total Customers
Europe	25,000	2.5%
North America	100,000	10.0%
Africa	350,000	35.0%
Asia	25,000	2.5%
Missing Values	500,000	50.0%

Figure 6.6 Frequency Distribution of Region

Gender	% of Records
Male	23%
Female	27%
Missing	50%
Income	**% of Records**
<25000	25%
25000–50000	25%
50000–75000	25%
75000+	23%
Missing	2%

Figure 6.7 Frequency Distribution of Gender and Income

Column Name: CREATIONDATE	Count	Percent	Cumulative Count	Cumulative Percent
[Missing]	1,822	3.4%	1,822	3.4%
2002	2,276	4.3%	4,098	7.7%
2003	2,027	3.8%	6,125	11.5%
2004	2,430	4.6%	8,555	16.1%
2005	3,213	6.0%	11,768	22.1%
2006	5,575	10.5%	17,343	32.6%
2007	6,699	12.6%	24,042	45.2%
2008	7,793	14.6%	31,835	59.8%
2009	7,239	13.6%	39,074	73.4%
2010	8,743	16.4%	47,817	89.8%
2011	5,418	10.2%	53,235	100.0%
Total	53,235	100.0%		

Figure 6.8 Frequency Distribution Report on Creation Date (Tenure)

in the form of analytical variables. The objective in this exercise is to have all meaningful information at the appropriate record level where any proposed solution will be acted on. In most cases, this would be the level of the customer or individual but is not necessarily confined to that level. For example, pricing models for auto insurance are actioned at the vehicle level, and retail analytics might focus on decisions that are actioned at the store level.

The data audit process should not only be conducted on new information or data sources but also on known data sources that serve as updated or refreshed information for a given analytical solution. Although the data audit process used in processing a new data source can be quite comprehensive, a less rigorous process may be used when assessing refreshed or updated data. This may be as simple as checking the number of records and fields that are received as well as some basic data audit reports on means or averages of key variables that are unique for a given organization or company. In any event, even a shortened version of a data audit report can establish a means of identifying problems or issues with the data used by analysts. An example of such a report is presented in figure 6.9.

In this example, key variables or potentially key business measures (KBM) are tracked over time to identify data issues or changes (both positive and negative) taking place in the business.

	Period 1	Period 2	Period 3
Record Count			
Avg. Purchase Amount			
Average Age			
Average Tenure			
Etc.			

Figure 6.9 Historical Perspective of Key Business Measures

Data audits, despite producing the least glamorous type of reporting information, are necessary prerequisites in any analytics process. Since all analysis starts with data, extreme diligence and respect for the data are required. This attitude to data helps to foster an appreciation of the many data nuances that can appear in a project. If analytics is going to be successful, data audits represent the first critical task in any given project

With this initial solid understanding of the data, the other critical component in creating the analytical file is establishing before and after (often referred to as pre- and post-) periods. In any data mining exercise, analysts must understand the metric being analyzed or targeted or modeled. This information is often referred to as the objective function or the behavior to be understood better. The following are some common business examples:

- How do we reduce attrition?
- How do we reduce credit loss?
- How do we optimize response?

In all these examples, analysts construct the analytical file in such a way as to create a solution to the above questions. But the behaviors or objective functions that are being analyzed ideally should be created in what is referred to as a post-period. Only the information used to create the objective function should be used in a post-period.

Meanwhile, the pre-period is used to create all the information that will be used to derive insights about the objective function or behavior in the post-period. The schematic presented in figure 6.10 best illustrates this. In this schematic, the analytical file is constructed where all the pre-period information occurs prior to the information in the post-period. In a way, analytics here are used in a more robust way because it considers

Pre period	Post Period
All behavior prior to Jan 01–2011	Jan 01–2011 / Mar 31–2011
All predictor variables	Objective function

Figure 6.10 Schematic of Pre- and Post-Period on the Analytical File

information (pre) that could be interpreted as predictive of something that will happen in the future (post).

Practitioners often assume that the objective function or post-period variable is relative straightforward and easy to create, which is not always true. Many examples exist in practice today. The first example is the use of proxies in creating a dependent variable. Requests are often submitted to build predictive models where no prior history exists. With no prior history, analysts cannot build a model that precisely predicts this behavior since it has not occurred yet. However, they can look for historical behavior that is similar to the behavior to be predicted. This similar type of behavior becomes the proxy for building the model. For example, analysts may want to build a model that predicts the likelihood of someone purchasing tennis shoes because the company has decided to promote these products. A reasonable proxy might be the likelihood of someone purchasing a tennis racquet because these products have been promoted in the past.

Another good example would be the purchase of squash products where none of these products have been promoted in the past. In this scenario, tennis products can serve as proxy or maybe the task can be seen on a broader context. Using tennis racquet purchase behavior as proxy could certainly be considered reasonable for purchases of tennis shoes, but when looking at a different sport additional analytics may be required for more insight. Correlation and affinity analysis may be conducted to find out how product purchase behavior aligns. For example, individual sports, such as running, golf, and tennis, may be more closely aligned with each other than with team sports, such as baseball, hockey, and football. Analysts can then identify the objective function as someone who is most likely to purchase a product related to an individual sport.

Building models to predict car purchase behavior can be a very challenging endeavor because car purchase behavior is infrequent with a typical average purchase cycle of five years. Building repurchase models is a daunting task because some customers will purchase a new car at three years, some at four years, some at five years, and the rest at six or more years. Analytics could be conducted to determine the time period of most of these purchases. In any case, the purchase period serving as the post period in defining the objective function encompasses a number of years. Creation of the analytical file would be based on all the characteristics and behaviors of customers prior to the post period. If the post period is three years, the information used is somewhat dated. A better approach to narrow down this post period window might be to use lease information; in that case analysts know when the lease expires. This expiration date can be used for the model and to determine who renewed a lease as opposed to those who did not renew. This information could then be used as a proxy for car purchase because it focuses on customers who are engaged with the car company. Using engagement by modeling on the basis of lease renewal represents a more practical approach to determining car purchase behavior.

Retention models often present the most difficult problems when trying to create the dependent variable. In most cases, there is not a precise behavior that defines cancellation such as renewing a subscription. Rather, cancel behavior is defined by inactivity in a specific period. As in the car example above, defining this period is not straightforward. For example, customer purchase inactivity of a week may define someone exhibiting cancel behavior in a grocery company, but the same period of inactivity to define cancels for tire purchases would be inappropriate. Analytics can be conducted to define more precisely what the purchase period should be.

Creation of the Analytical File and Pivot Table

One potential end deliverable in creating the analytical file is the creation of a pivot table. To create the pivot tables, further summarization is done based on how the data is to be viewed against a given key business measure (KBM) or set of KBMs. The complexity of this summarization will increase as the number of dimensions and the range of outcomes within a given dimension increase. For example, if the average spending for a given business initiative is assessed in terms of gender (male and female) and education level (college-educated and not college-educated), then there are 4 (2 x 2) possible views or

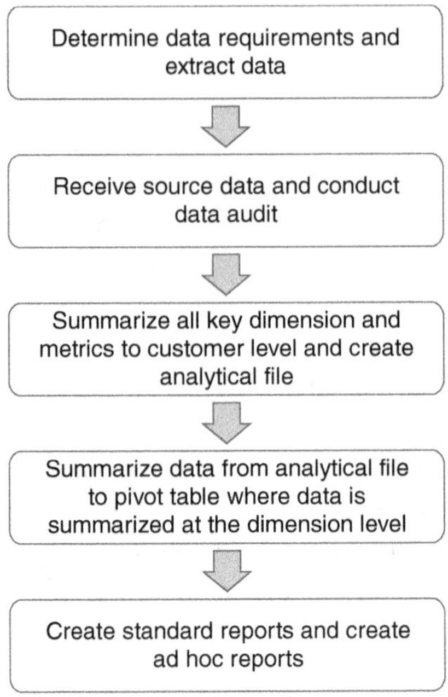

Figure 6.11 Flow Chart for Creating Business Reports from Pivot Tables

dimensions. However, if spending is analyzed by gender, education level, age category (<25, 26–35, 36–45, 46–54, 55+) and by household size category (1, 2, 3, 4, 5+), then the number of views or ways that spending may be looked at explodes to 100 (2 x 2 x 5 x 5) possible views or dimensions. The pivot table represents the source file in terms of creating the desired measurement reports. A schematic as an example of this process in terms of the original data sources and the measurement reports is shown in figure 6.11.

The development of measurement reports requires much planning in terms of what is being measured. At the same time, a deep understanding of the data environment is absolutely critical to implementing any type of measurement plan. This requires a team-oriented approach between marketers and the data analytics group. Organizations with a strong team philosophy uniting marketers and the data group will have reporting systems that are most effective in the measurement of ROI.

The Value of Understanding the Data Environment: Case Study of a Retailer

The retailer in the example wanted to conduct more CRM programs and realized that data was the foundation of success. The objectives were twofold:

1. Understand the data environment
2. With this understanding, develop a best customer program.

For objective 1, a standard data audit program was conducted. In addition to dumping 10 records, diagnostics were also created that indicate the total number of records in each file that were loaded into the system. This was done to ensure that whatever is received from IT is identical to what is loaded into the system.

Once this data was loaded into the system, standard data audit diagnostics were completed on all files. A frequency distribution is shown in figure 6.12.

This report revealed that 50% of customers had no known address. In other words, any personalized contact to these customers requiring address would be extremely limited. Of customers with a known address, 70% resided in Ontario. Both these findings provided learning that (a) direct marketing or any CRM system would be severely limited with 50% of the customers having a missing address and/or postal code, and (b) most activity seems to be occurring in Ontario.

Region	# of Customers	% of Customers
Prairie Provinces	25M	2.50%
Quebec	100M	10%
Ontario	350M	35%
West	25M	2.50%
[Missing Values]	500M	50%
Total	1MM	100%

Figure 6.12 Frequency Distribution of Region

Last Name	First Name	Address	Postal code	Phone #	Customer ID
Boire	Richard	123 Main Street	L1W3K6	905–489–2124	1000123
Boire	Richard	123 Main Street	L1W3K6	905–489–2124	1000126
Boire	Richard	123 Main Street	L1W3K6	905–489–2124	1000129
Boire	Richard	123 Main Street	L1W3K6	905–489–2124	1000123

Figure 6.13 Example of Same Customer Mapping to Separate Customer IDs

Another component of the data audit exercise was to match the two files together in order to determine the quality and integrity of the supposed match key. The understanding was that customer number or ID was the unique qualifier for each customer. In fact, this customer ID was not only unique, it was too unique. Primary forensics involved matching the customer file to the billing file. Here the usual outcome was produced, and one customer ID matched many customer IDs in the transaction file. This outcome is expected as customer will have had several transactions with a given company and thus a one-to-many relationship. Further forensic investigation on this file and the dumps of records revealed the scenario shown in the example in figure 6.13.

This outcome indicated that a given phone number was linked to multiple customer IDs. Perhaps there may be more than one person in a household with the same phone number who are separate customers. However, deeper investigation of the data was conducted by examining name and address of these records, and it was discovered that duplicate phone numbers did not indeed map to separate customers but rather to the same customer (the Richard Boire example above matches to 4 unique customer IDs). In these cases, the same person was mapped to multiple IDs. In presenting this information to the company, it was decided that a better understanding of the process of how a customer obtains a given ID was required. In the investigation, analysts discovered that customers obtain a new ID if it is their first time in a given store. This led to the discovery that customer IDs were not unique to a given person but to the person and the store. In other words, if a person conducted business at five different stores, that customer would have five separate IDs. In the example, Richard Boire has gone to three separate stores (1000123, 1000126, and 1000129) and has shopped twice in the same store (1000123).

Analysts then realized that the phone number would be the optimum match key. Each store was required to capture a customer's phone number every time the customer made a purchase. Specific recommendations included the following:

1. Use phone number as the basis to build unique customer keys.
2. Use software that can append name and address to customers with missing values in these fields. With phone number available on 100% of the records, this is the critical key in a table provided by the software that contains fields mapping phone number to name and address.

3. Consider programs that expand the retailer's presence beyond Ontario.

The key consideration here is that the better insight into the existing data environment allowed us to identify and adopt these data strategies.

The second component of the research dealt with conducting analysis. In discussion with this retailer, it was discovered that management did not know who the company's most valuable customers were. Were 80% of revenues being driven by 20% of the customers or was it more like a 60:40 ratio? The basic Pareto rule regarding customers and revenue contribution was completely unknown to management. In identifying how revenues were distributed among customers, the company wanted to better understand the data and to create a profile of its best customers. Accordingly, data was extracted that would provide the necessary solution. The extraction focused on active customers only (i.e., customers who had purchase activity in the past year). This led to the first key finding: half the retailer's customer base was inactive, and this presented a significant future business challenge but one that was outside the scope of this initial project. Figure 6.14 presents a decile table ranking customers by total purchase amount, which was agreed to by the retailer and used as a proxy for value.

Through this analysis of customers, close to 60% of revenues were contributed by 20% of the customers; consensus coalesced around the notion of creating a best customer program for this 20% (high value) group. The marketing department also wanted to better understand what these customers were like.

As shown in figure 6.15, compiling information about where these customers came from and what products they purchased in the past provided a cursory profile.

Segment	% of Customer	Avg. Annual Sales Last Year	% Contribution to Total Sales by Interval
High Value	0–5%	250	25%
High Value	5–10%	170	17%
High Value	10–15%	110	11%
High Value	15–20%	70	7%
Low Value	20–25%	50	5%
Low Value	25–100%	25	37%

Figure 6.14 Example of Value Segmentation Report for Retailer

Product Type	% of All Customers	% of All High-Value Customers
Bought Product A	20%	40%
Bought Product B	30%	10%
Bought Product C	25%	25%
Bought Product D	25%	25%
Total	**100%**	**100%**

Region	# of Customers	% of All High-Value Customers within Reported Postal Codes
Prairie Provinces	5%	5%
Quebec	20%	30%
Ontario	70%	55%
West	5%	10%
Total	**100%**	**100%**

Figure 6.15 Examples of High-Value Profile Reports

Customer Segment	Control	Test	Do Not Promote
High Value/ Best Customer	175,000	5,000	5,000
Balance	5,000	5,000	5,000

Figure 6.16 Example of Simple Test Matrix for Best Customer Program

- Best customers tend to reside in Quebec and the West
- Best customers tend to buy product A

With this information, the next step was to act on learning in a best customer program. A list of names for this program (200,000) had to be created. In creating this list, a "do not promote" list was also created and a list of names that were promoted to but were not part of the best customer group. A schematic of this list selection is shown in figure 6.16.

The names (175,000) are all front-loaded as the segment (best customer) and communication strategy (control) are expected to be the winning strategy. The remaining cells are about learning:

- Did the campaign yield incremental purchase results over and above those of customers who received no campaign offer?
- Which communication strategy works best (test vs. control)?

- Do higher value segments (best customer) yield more incremental results than lower value segments (balance)?

The program was launched and easily exceeded the company's expectations and thus justified the shift from mass-marketing to CRM marketing.

This page intentionally left blank

CHAPTER 7

DATA MINING PROCESS: CREATION OF THE ANALYTICAL FILE WITH EXTERNAL DATA SOURCES

TODAY, VIRTUALLY ALL ORGANIZATIONS CONDUCTING ACTIVITIES IN data mining purchase external data sources. These data sources represent information a given company is unable to collect solely on the basis of its internal activities. Another perspective on this is that this type of data represents information that is available to the public either free or at some cost. The level of data mining sophistication will determine the extent of these external data source purchases. The reason for purchasing this data is to augment existing information about customers. How can this external data augment a company's data about individual customers (especially when most of this external data is at a less granular or a more aggregate level, such as data about postal walks or census information)? This can be better understood if the kind of data available is considered at the level of the individual customer. For the most part, a very substantial portion of this data concerns transactions or purchase behavior. External data sources that are valuable to most organizations deal primarily with demographics that are in limited supply in an organization's own database. For example, demographic variables at an individual level, such as age, income, and gender, may be missing more than 50% of their values when researchers try to extract this information from a customer database. The use of external overlay data can provide information related to these characteristics based on where the individual lives.

USING EXTERNAL DATA FOR EXISTING CUSTOMER PROGRAMS

First of all, purchasers must determine whether the data is needed for acquisition programs or for existing customer programs. For existing customer programs, most organizations in today's CRM-driven environment have databases of existing customers and their purchases. The organization and structure of this data may vary from company to company depending on the level of sophistication in database marketing. Yet, regardless of how this data is structured and organized, for data mining purposes, the data is at an individual level. Data miners will always strive to use as much data on the individual level as possible because this will always provide superior results than the more aggregate data captured at a geographic level. But these results can be compromised if there are quality issues with the individual-level data. For example, age and income might be reported on an individual level. Yet, if over 90% of customer records are missing the values in these fields, this individual level data on the remaining 10% of customers is not going to be very useful. In this case, businesses should look at aggregate data sources, such as Statistics Canada and specifically Statistics Canada census data or the appropriate external data provider in their country. Aggregate data means that customers residing in the same postal area would have the same Statistics Canada values, and customers residing in different postal areas would have different values. Appending aggregate-level data (Stats Can census area) to this customer file would at least provide complete age and income information for 90% of records that currently do not have any of this information at all. Using this aggregated data for income and age would yield superior results compared to the status quo of individual level data that has 90% of this information missing.

In addition to enhancing information where there are data quality issues, aggregate-level data can also provide breadth of information. For example, areas such as ethnicity, occupation, religion, education, and a range of other Stats Can demographic information are unlikely to be directly available on any customer database. Appending this type of information to an existing customer database can add some value to a modeling or profiling exercise. However, the rich individual-level information related to the purchase or transaction information of the customer will produce the stronger modeling/profiling variables relevant for targeting purposes. This aggregated demographic information can also have some impact in providing more general insights that can be used to develop better communication strategies. For example, the demographic and geographic profile of a certain group of customers may indicate that these customers live predominantly in areas with high income, high number of immigrants, and a large percentage of self-employed people. Certainly, marketing executives might want to develop unique communication strategies for these groups rather than offering them the standard generic promotion of their product or service.

The bigger impact of external data will be seen in its use in acquisition programs. In fact, data suppliers market themselves more on the overall impact on acquiring new customers. External data is much more needed in building any data mining solution for acquisition programs rather than adding information to existing customer programs. Typically, name, address, and postal code represent the available pieces of data for any acquisition program. Postal code is the key link in being able to append all the Stats Can demographic data to name and address records. Stats Can data is offered in two main types of products. The first product is Stats Can tax filer data. The data here is organized based on postal walks and represents approximately 800 households; the information it contains is compiled from annual tax returns. Such data would contain income-related information, such as income earned from employment, investment income, charitable deductions, etc. The second file, Stats Can census data, contains information based on enumeration areas for approximately 400 to 500 households. Although this file contains some income data, it lacks the other measures of wealth that are contained in the tax filer data. However, this second file is much richer in demographic information and contains information pertaining to ethnicity, religion, language, education, occupation, etc. Furthermore, it is much more granular as data is aggregated for every 400 households as opposed to 800 households, which is the case with the tax filer data. A limitation of this census data is that it is updated only every five years, which is when the Stats Can census survey is conducted. It is important to remember that in both cases (Stats Can census and Stats Can taxfiler) the data is collected at an individual level but is then summarized to the appropriate geographic level (postal walk for taxfiler data and enumeration area for census data). It is this summarized data that is available to data miners; in this way the data collection and use adheres to the guiding principles and regulation behind any current privacy legislation.

In Canada, recent decisions by the federal government have now made the gathering of Stats Can census data by the Canadian population a voluntary exercise instead of the more robust mandatory exercises used in all previous

Stats Can surveys. Stats Can taxfiler data, though, does not have this same limitation as Canadians are compelled to file annual income tax returns.

In 2011, the census data was gathered in this fashion, which makes comparisons between periods virtually meaningless. For social demographers, their work now encompasses more limitations in trying to identify key social trends and behaviors that rely on historical census behavior over a number of different time periods. In data mining, the implications of the new voluntary approach have less impact. Most data mining objectives are less about comparing demographic trends over time and more about rank ordering or differentiation of records. By using the data at a point in time, data miners care less about how the data was collected and more about its richness and breadth in being able to create new variables that achieve objectives of being able to rank order and differentiate records.

What to Consider When Purchasing External Data for Data Mining Purposes

Purchase decisions regarding various data sources can become extremely important in any data mining exercise. In purchasing data for data mining purposes, the data is not used to replace existing data but rather as additional data sources supplementing the existing data environment. However, it has still not been determined whether or not this new data can provide significantly better results over data already in the existing data environment. The use of lift charts, which will be discussed at much greater length throughout the remainder of the book, represents one option. For example, the incremental benefit of such charts (see figure 7.1) is the additional rank-ordering or incremental lift provided by the external data if it is used for a predictive model or targeting solution.

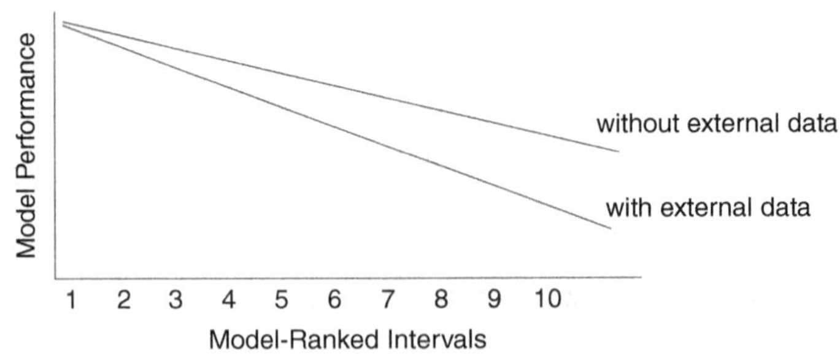

Figure 7.1 Lift Charts with and without External Data

CHAPTER 8

DATA STORAGE AND SECURITY

IT IS OFTEN THOUGHT THAT THE MAIN REASON FOR THE PROMINENCE OF data mining in today's business world is software and the specific mathematical/statistical applications embedded in these tools. However, many of these tools have been in existence for over 25 years. The underlying reason for the historical lack of use in the business world has been expense. The costs involved in providing the appropriate computer resources to run these applications were simply exorbitant. In today's technical environment, the limitation is not necessarily on the processing side but rather on the ability of human beings to process this data into meaningful information. In fact, the use of technology such as Hadoop and MapReduce from organizations such as Apache has even accelerated this data processing capability to a new level. The world of big data has resulted in the type of technology that allows us to deal with the three Vs of Big Data: large amounts of data (volume) in different formats (variety) in many cases are being streamed in real time (velocity).

These technological advances have led to wider use of data mining techniques because of two factors:

- Increased ability to store and access data
- Increased ability to process instructions

The increased ability to store, access, and process data is fundamentally a function of the significant advancements and reduced costs of both computer memory and disk drive storage technologies. The improved availability/affordability of computer memory (RAM) facilitates the temporary storage of much larger data sets and allows more and faster processing (reads, writes, calculations) to be conducted with data that resides in memory rather than data that resides on permanent storage devices (disk drives). The improvements in permanent storage devices (disk drives) have allowed online storage of much larger quantities of data at reduced costs by volume and have provided faster read and write capability and greater reliability, thus allowing much more data to be stored in an online (readily accessible) state. From a business standpoint, this has translated into the ability to access large amounts of data at ever-increasing speeds.

As data represents the primary tool of the data miner, the security of this data or information as outlined in legislation is of great interest to the data miner. The protection of the data entrusted to data miners is not an option but an obligation. Data miners are constantly exposed to the transfer of data either in receiving or sending data. The data miner should be aware of the sensitivity of the information contained in the data. This sensitivity will dictate the various protocols and procedures required in any data transfer. Expectations regarding the handling of data have changed considerably. Currently, sensitivities regarding data theft and loss have contributed to more stringent protocols for data transfer. Encryption techniques and secure FTP (SFTP) transfer protocols are now the norm.

Another mechanism for ensuring that data remains secure is to have proper data retention policies. Data miners, as a rule, would like to keep data as long as possible and have it readily accessible to them. But there is always a certain degree of risk in keeping data for extended periods of time, especially data containing names, addresses, and other personal information. Because data miners deal with data that is used for creating and applying a solution, they also need this data when validating the solution that has been applied to a particular business initiative or marketing campaign. For instance, the name and address information of a direct mail marketing campaign using an applied data mining solution is required in order to link it to back to the responders of that campaign, which would also contain name and address information. The rule of thumb in most of these cases is usually that data will be kept for six months; this means that the expectation is that a back-end or performance analysis should be done within six months of the campaign launch. If files are older than six months, procedures should be in place to archive this data to offline storage or destroy the data as per the retention policies of the data owner.

CHAPTER 9

PRIVACY CONCERNS REGARDING THE USE OF DATA

PRIVACY IS A TOPIC WHERE PEOPLE WILL HAVE STRONG OPINIONS. IN Canada, much of this opinion has been formulated into many of the principles and guidelines outlined in PIPEDA (Personal Information Protection and Electronics Document Act). The legislation in many cases is very clear on what marketers and data miners can and cannot do. But there are some grey areas when terms such as reasonableness are used. For the most part, marketers and data miners respect the legislation since it encapsulates what has been best practice in many organizations for many years. It is simply a good business practice to be respectful of and attentive to the privacy needs of consumers when attempting to properly market the right services and products to them.

The industry's practices and disciplines will result in legislation that could be very tolerant or very restrictive. The legislative direction within this area will ultimately be determined by how respectful marketers are regarding consumers' rights to privacy. This implies that data miners as well as database marketers must be proactive on privacy issues.

Although there are 10 guiding principles within legislation, 3 impact the area of data mining the most:

- Identifying purposes or use of information
- Consent
- Security or safeguards

The principle of security or safeguards was discussed at length in the preceding chapter so we will focus on the first two principles here in this chapter.

In many ways, using data mining on customer data is not as contentious in terms of privacy as might be expected by some privacy experts. Why? With mathematics and science, business users can make decisions on groups and not individuals. The reality of any data mining analysis is that business users will apply data mining results at an individual level; however, the insights and learning regarding these decisions are based on the results and performance of a group of individuals.

IDENTIFYING THE PURPOSE

Consent clauses should clearly state the purpose of the consent. For example, if information is collected for the purpose of either renting or selling names to a third party for marketing activities, then this should be clearly stated in the clause. If information is collected to help sell other type of services and products to customers of the same organization, then that also should be clearly stated. However, some privacy advocates might argue that more details should be provided to consumers on how information about them will be used. For example, some people believe that consumers should be informed that mathematical and statistical techniques will be used to help analyze the information about them.

Should marketers really have to explain the intricacies of multiple regression and multivariate techniques to the public at large? This would be an arduous task, but maybe it should be done simply because it is the right thing to do. How do we decide what is right? Regarding this legislation on data use, "right" is conceived of in terms of what is reasonable. Is it reasonable for consumers to know precisely how marketers are analyzing their data? What is reasonable is that consumers want to know how marketers will use the information in regard to other marketing activities. Consumers also expect that future marketing using this information would be related to the initial marketing activity in which they gave their consent. For example, it is reasonable to expect that a credit card customer who has consented to provide information would be offered insurance to protect his or her credit rating in case of a job loss. However, it is unreasonable for donors of a nonprofit company to expect receiving promotions for new credit cards.

This does not mean that nonprofit organizations are restricted in their activities, but to highlight that all organizations must consider what is reasonable from the customer's viewpoint.

Reasonableness is not clearly defined, and the test for reasonableness pervades the legislation. This is why many pundits believe that this legislation is constantly evolving, especially in the digital information age. All involved in using this legislation will gain greater knowledge and insights with more practical experience in the application of the law.

Do consumers really care about the use of data mining tools and statistics to analyze data? If these tools are used to select consumers, then it might be argued that consumers do not care about how they were selected but only *that* they were selected for particular services and products. The use of these tools in selecting certain individuals is of no interest to those concerned as long as they understand they could be selected for a particular offer or promotion. This might be the threshold of reasonableness today for the consumer. As stated earlier, with more experience in the application of the law, the threshold of reasonableness might change, and consumers may then have to be informed about how their information will be analyzed. If that is the case, then the book *Statistics for Dummies* will become a best seller.

CONSENT

Consent can seem relatively straightforward in terms of whether or not the customer gave permission for use of his or her information. However, this principle is often more complex since it ties in very closely with how a given company will use the information. For instance, a company may promote products and services to its existing customers as opposed to renting or selling its existing customers' names to a third party; these are very distinct business activities. Some people argue that the former activity of continued marketing promotion to existing customers requires fewer restrictions on obtaining customer consent than would be required for renting or selling the customers' names to a third party. In fact, even regarding renting or selling names to a third party, there may be different levels of customer consent. For example, renting names to a business publication would be very different from renting a name to a charity. However, in many cases renting and selling of names is no longer an acceptable business practice. For companies that continue this practice, the consent requirement might be very stringent, and this usually means that the company requires opt-in consent as opposed to an opt-out decision. Other restrictions might also require disclosure of the specific organizations that would be allowed to receive the customer's name and address.

Opt-In versus Opt-Out

The distinction between opt-in and opt-out consent decisions is very important in terms of gaining customer consent. To opt in, customers have to initiate some activity before their consent is deemed to be obtained. For example, often they must check or click a box in a promotional offer or communication piece from a company. Only when the customer has checked, clicked, or filled in the check box, has consent been given. A blank check box is not deemed customer consent. In other cases, questionnaires are used to obtain additional customer information, and they might contain a clause stating that filling in the questionnaire implies consent to the use of the questionnaire and any behavioral information for future marketing purposes.

From a marketing standpoint, opt-in consent is very restrictive for most organizations and would severely impact their ability to conduct their business profitably. Many people will do nothing to initiate receiving promotions. This does not mean that they do not want to receive additional products or services but simply that they either don't care about filling in the check box or have not paid attention to the offer. With opt-in consent, the net number eligible prospects available for marketing would be significantly smaller, and then many marketers would not be able to take advantage of the economies of scale a large volume would offer. Many such companies would cease activities that in the past have been very profitable and also provided value and service to consumers. With these no longer profitable activities jobs would also be lost as well as additional sales dollars, and this would be detrimental to our economy. However, in digital marketing, due to the proliferation of messages bombarding customers, the opt-in scenario is often the norm. In fact, recent anti-spam legislation by the Canadian government now requires e-mail marketers to adopt the opt-in rules of consent.

Because of the deleterious effect of opt-in clauses, many organizations have attempted to deal with obtaining consent by using the so-called opt-out provision where permitted. In the opt-out provision, the customer has to fill in a check box in order to refuse consent to the gathering and use of his or her information. A blank response to this check box is considered implied customer consent. The key for companies choosing to use opt-out consent is to make this consent opt-out clause clearly visible and display it prominently in the overall text. Any attempt to bury the clause in small print and in the middle or near the end of the text defeats the purpose of the overall privacy legislation. Furthermore, for organizations priding themselves on being marketing experts, clearly defined opt-out clauses simply represent the best business practices that will help to further cement their relationships with their customers. Keep in mind that these practices are more relevant for nondigital channels such as direct mail.

Channel Consideration

Regarding customer consent, another key consideration is the vehicle or channel used to promote products and services. For example, direct mail, e-mail, and outbound telemarketing calls may not all require the same level of customer consent. Perhaps different thresholds should be used depending on the channel. Some channels are more intrusive than others. For example, receiving a direct mail piece is less intrusive than receiving a phone call, especially if it is at supper time. For marketers, opt-in consent may be more appropriate for digital and outbound telemarketing channels while the less restrictive opt-out consent may be more appropriate for direct mail.

Time Period for Consumer Consent Decisions

Suppose a consumer chooses not to give consent to future promotions and/or offers. One of the areas that the legislation does not address is the period of time for which the consumer's decision should be honored. Should a name forever be taken off the marketing list once that customer has refused consent? Marketers realize that this "do not promote" file will grow over time and at some point represent an opportunity. This is similar to an organization's list of lapsed customers; customers having had no activity with the company for a certain period of time will be targeted by marketers in an attempt to restore the lapsed customer's activity. For the list of "do not promote" names, the logic works the same way: after some period of time, it would be reasonable to reassess the status of these customers. Perhaps after a year, a promotion specifically geared to reassess the status of these customers might be tried. An opt-in statement could also be presented on the customers' monthly billing statement if they ever changed their mind and made a purchase.

Likewise, when a person gives consent, the question is how long this would reasonably be in force. The legislation doesn't address the issue of period of time for both consent and refusal of consent. This issue also will develop as more experience is gained.

All these examples refer to customers, but what about prospects? Suppose a company sources its names from the telephone book and as part of its overall acquisition strategy decides to build a prospect history database. As the database becomes populated, the marketers run acquisition campaigns using this database. Over time, they collect information related to the type of promotion, frequency of promotion, and recency of promotion at the level of individual prospects. This information is tremendously valuable because organizations can segment prospects based on promotion frequency and recency. Are organizations respecting consumers' right to privacy? One could argue this both ways. Certainly, consumers would not know that this type of information is used to target them. But let's consider the information being used in more detail. The information collected relates to what the company has done to the prospect in terms of previous promotions. This type of information is company-driven rather than customer-driven. There is no customer-driven information here relating to specific consumer activities or consumers' transactional behavior. From this perspective, this information about prospect history could be considered somewhat less sensitive since no customer-driven information is contained in the database. Given the less sensitive nature of this information, marketers could use it to better target their prospects and still not be intrusive.

In social media marketing, much of customers' information would be deemed of a prospect nature as well. Yet, the proliferation of this medium has resulted in new areas of analytics to leverage information such as text mining. People generally would not expect that this information could and would be used by a company for marketing purposes. Yet, in this case, behavior of the consumer as opposed to that of the company is captured without the consumer's knowledge. Thus, it is questionable whether or not consumer privacy is respected here.

Renting or Exchanging Names

Data miners are often responsible for sourcing new names or prospects in order to acquire more customers. For instance, when consumers subscribe to certain magazines, they may see a clause telling them that their name might be rented or traded to other organizations. The clause would then require an opt-out check box so subscribers can check the box if they do not want their name to be traded or rented to a third party. In other words, subscribers or customers must be proactive in deciding whether to allow their name not to be exchanged or rented to other organizations. But what if a magazine such as *Playboy* decided to exchange or rent its names to another organization? Would a simple opt-out clause be sufficient in this situation? One might easily argue that opt-in consent is the only acceptable type because the sense of what is reasonable (always very subjective in the context of law) is raised to a new level due the information being more sensitive than with other types of magazines. In this case, consumers must directly communicate to the company that they want their name to be exchanged or rented to another company. Doing nothing in this case implies that the consumer has not given consent. In this opt-in scenario, consent must be explicit through some action by the consumer whereas in the opt-out scenario, consent is implicit with no action by the consumer.

Although the example of *Playboy* magazine represents an extreme scenario, data miners are often more involved in using customer information to promote other types of products and services to existing customers. For example, suppose a credit card company wants to sell creditor insurance to its customers. This product offers credit card balance protection to a customer in times of economic difficulties, such as a job loss. In these situations, the insurance company continues to pay the required minimum payments on any outstanding credit card balances. Before offering this product, the company would send an opt-out clause to customers that allows them to opt out of all communication about offers. This clause would probably also mention customers' information is used to help offer them the best services and products. If customers do nothing, it is assumed that consent is given. The company would then use the credit card activity and information of all customers who have not opted out. The appropriate names would then be selected based on certain customer information regarding their likelihood to respond to an insurance product and their likelihood to be profitable.

USING CREDIT RISK DATA

Credit risk data presents excellent information for targeting customers for marketing purposes. In many marketing models, this was often one of the strongest variables. Consumers expect that this information will be used for making credit decisions because this is clearly stated on the credit application consumers fill out.

However, it's reasonable to ask whether consumers expect that this information is also used for marketing purposes. Clearly, the test of reasonability would fail here, and credit risk data is not generally used for marketing purposes at this time. However, organizations such as Equifax, which understand the value of data, have created data products containing this credit-related information but at a postal code level or zip code level. Organizations purchase this data as another source of geographic information that can be overlaid onto their database. Although the geographic or postal code/zip code credit risk data is not as powerful as credit-risk data about individual customers, it can still provide another piece of valuable information to help target customers for various products and services.

In conclusion, regarding privacy, it is important to understand that there are no real experts in this area since consumer privacy is an ongoing and evolving concern. Even lawyers, though well-versed in the discipline of law, do not have the deep understanding of database marketing and analytics that industry experts have. As a result, marketers are dealing with legislation designed to protect consumers but without the lawyers necessarily understanding the practical business implications or the data and technology used to solve these problems. However, over time, lawyers, businesses, and the data mining community will gain enough business experience to improve the legislation in the interests of both businesses and consumers.

This page intentionally left blank

CHAPTER 10

TYPES AND QUALITY OF DATA

DIFFERENCES BETWEEN DATA LEVELS

If there is one opinion of data mining that experts almost universally agree with, it is that the data components and content are the core of data mining success. Despite the latest advancements in technology and software, which all purport to significantly improve data mining results, data is clearly the driver behind any successful data mining project. Having said that, it is important to understand that there are different types and levels of data. Knowledge of these types and levels of data can provide insight into how performance will be impacted.

Data miners tend to look at the quality of data in regard to four main areas:

- Data granularity
- Biasness of data
- Data coverage
- Range of data values

DATA GRANULARITY

In any data mining exercise, the objective is to acquire as much individual-level data as possible. Since solutions will always be applied to individuals, development of a solution at this level allows analysts to utilize unique values of a field for each record. An example is provided in figure 10.1.

What are some preliminary observations we can make regarding these six customers even without experience in statistics?

Customer	Age	Income	Response
1	75	80K	Yes
2	28	40K	No
3	45	60K	No
4	33	45K	No
5	62	70K	Yes
6	65	75K	Yes

Figure 10.1 Example of Customer/Individual-level Demographics

- Older customers are more likely to respond
- There is a relationship between age and income where higher age translates into higher income
- Higher income customers are more likely to respond

Now, suppose we have geographical or aggregated data (namely, Statistics Canada demographics at the census level; keep in mind that this applies also to many other countries as demographic data would also be collected there at the appropriate geographic level). Typically, this consists of aggregate-level data at the level of enumeration areas that includes information such as income, age, gender, education, ethnicity, occupation, language, mobility, etc.

This information is typically appended to the customer file by postal code through the use of a conversion table containing both postal code and enumeration area. Customers residing in different enumeration areas will have different Stats Can values. However, customers residing in the same enumeration area but in different postal codes will have identical Stats Can census values. In appending data, data miners will always attempt to obtain the most granular type of data. This implies that data summarization to a given level is minimized. For example, appended data could be available at a variety of levels:

- Forward Sortation Area (FSA-1st 3 digits of postal code) level: approx.10,000 households
- Postal walk route level: approx. 800 households
- Census/enumeration area: approx. 400 households

It is obvious that the census data would be the most attractive option given our objective of granularity. However, the postal walk data contains financial information not available at the census level. This includes data such as:

- Charitable donations
- Registered Retirement Savings Program(RRSP) room and contribution
- Dividend and investment income

Postal walk information is also updated every year, but the census data is updated only every five years. Depending on the objective of the data mining project, different purchase decisions in terms of external data will be made regarding postal walk versus census data. Outside Canada, this approach to making these type of decisions still applies.

In order to appreciate the loss of information when using external aggregate-level data, let's take a look at how this data might appear if the individual level data existed for six customers on age, income, and response along with the census level data. Figure 10.2 provides a table for these six customers showing clear trends: responders are much older and have higher income.

Let's say that now this information is only available at the census level; it would then be summarized as shown in figure 10.3.

By comparing the two tables in the figures, we can see that our findings regarding trends differ when we look at each table separately. For instance, the above findings at the individual level are not as pronounced when summarized at a census level. The relationship between age and response is very minimal, and the relationship between income and response is also very minor.

Census Level	Age	Income	Response
10010001	75	80K	Yes
10010001	28	40K	No
10010001	45	60K	No
10080002	33	45K	No
10080002	62	70K	Yes
10080002	65	75K	Yes

Figure 10.2 Example of Six Customer-Level Records with Census-Level Data

Census Level	Age	Income	Response
10010001	49.3	60K	0.33
10080002	53.3	63.3K	0.66

Figure 10.3 Example of Six Customer-Level Records rolled up to Census level

These simple examples illustrate the point of how information is lost when aggregating data to a certain level. Keeping data at its most granular level will always be a paramount objective for data miners.

Biasness of Data

One of the fundamental questions that must be asked at the beginning of any data mining exercise is whether the data is representative of how the solution is going to be applied. What is meant by this? A huge consumer survey has been done, and the results and learning will be applied to the Canadian population. It was also discovered that most of the respondents of this survey are female (approx. 80%). Applying any learning from this sample would be erroneous since the usual gender split in the population is about 50/50. Good data miners are able to work with the data they have and arrive at some solution. In this case, data miners might stratify the above consumer survey sample in order to obtain an analytical sample that does have a 50/50 gender split. This would involve the random selection of every fourth female survey responder to be included in the survey.

Another helpful example is the application of solutions developed at a national level that are then applied regionally. Without getting into any political debate on the issue, it is understood that Quebec is indeed different from the rest of the country, primarily because of its official language of French. Seasoned direct marketers will say that what works in the rest of the country does not work in Quebec. This suggests that solutions for Quebec must be developed solely from information gathered in that province. In fact, in my experience applying solutions developed in Quebec to French-speaking citizens outside Quebec is not usually optimal. A similar phenomenon is observed when applying solutions for the Rest of Canada (ROC) to English-speaking Quebec residents. In other countries, the same regional disparities must be considered when applying data mining solutions. For example, are solutions that are appropriate in Texas really appropriate in New England?

These two examples illustrate that some basic demographics need to be considered in evaluating whether or not there is a bias in the data. Aside from region and gender, other characteristics to consider are income, age, and household size. In trying to determine the biasness of a given sample, a simple spreadsheet exercise can be produced that displays the basic statistics (means) between the Canadian population and the sample being considered for the data mining project; an example is provided in figure 10.4.

Demographics	Canadian Population	Sample for Data Mining Project
Live in West	20%	19%
Live in Quebec	28%	27%
Live in Ontario	40%	41%
Live in Maritimes	12%	13%
Income	$38,000	$37,000
Age	45	44
% Female	51%	49%
# in household	1.5	3

Figure 10.4 Comparison of Population vs. Sample Averages

From these results, it is evident that the sample has a large bias toward larger families. Applying any solution from this sample would imply that its impact would be greatest on larger families.

Bias of this type is not necessarily bad if the biased nature of the universe is understood before a data mining solution is applied. Understanding how different or biased the data environment is between the time when the solution was developed and the time when it is applied is important. By comprehending this difference, it is possible to select various alternative courses of action, such as reducing the performance expectation of a given solution or developing a new, more optimal solution for this biased universe.

Data Coverage

Particular files or databases contain many fields, but this does not mean that the information is useful. It is not uncommon for database environments to have many fields but few recorded values in any of these fields. Dealing with missing values or data fields where most of the records have no recorded value is one of the most common tasks data miners have to undertake. The use of a data audit (previously discussed) allows analysts to uncover how prevalent the missing value situation is for every given data field. With frequency distributions, it is possible to obtain a very good sense of the usefulness of a variable based on the extent of missing values. A useful rule of thumb is that any variable or data field with more than 90% of the records having no recorded value would not be considered a worthy variable in any data mining analysis. Organizations can use the results of this component of the data audit to help create a standard list of variables that can serve as a template for all future data mining exercises.

Range of Data Values

Examination of the range of values for variables allows the data miner to identify potential outlier issues particularly for variables that are continuous in nature. Determining the number of unique values also provides insights on what data miners may need to do in terms of manipulating the data. For example, if the entire analytical file represents only males, there would only be one value for gender. Gender in this case would be meaningless in any future data mining exercise because there is no other outcome that can be analyzed for comparative purposes. In some cases, character variables such as product codes may comprise hundreds of different outcomes. This kind of scenario would indicate to data miners that these outcomes must be grouped or categorized rather than be limited to binary variables for each outcome. An even better example of this is how data is analyzed geographically. Analysis by geography tends to be done at the level of region or province. Analysis of postal codes or zip codes does not provide enough mass in terms of who lives in the postal /zip code (very few) relative to those who do not live in the postal /zip code sample (everybody else in the sample).

Data Enhancement Services

The business of supplying data has grown significantly over the past 20 years. This growth can be attributed directly to the growth in data mining. Historically, clusters were used solely as a means to target prospects. But as the level of sophistication in using data has increased, analysts are not only looking at the clusters but also at the raw source data that was used in building these clusters. The analytical mind-set is to consider a variety of data inputs when building data mining solutions. With this kind of mind-set, data suppliers themselves have become more sophisticated and offer data enhancement products beyond just cluster codes. One company offers data enhancement products that provide increased targeting capabilities among different ethnic groups. Another organization offers demographic postal area information that goes beyond the information provided by Stats Can census and tax filer data. In fact, this organization uses both these sources as well as a variety of other data sources to provide annual demographic data at a postal code level.

Although this data is much more granular (postal code level) and would therefore appear to be superior to the more aggregated Stats Can data, this postal code demographic data represents estimates and not raw or source data. Estimates are based on some form of mathematics and are going to have some degree of error. Yet, in building data mining solutions for differentiating among customers and/or prospects based on a desired behavior, this type of data can indeed provide more powerful inputs to a data mining solution than the raw Stats Can data. But again, data miners will always strive to consider using both the derived granular estimates as well as the more aggregate raw source Stats Can data as data inputs in any data mining solution. However, it is extremely important that data miners understand what is source data and what is estimated data when they build a given solution. Furthermore, understanding the level of granularity (i.e., postal code vs. enumeration area vs. postal walk vs. FSA) provides additional insight into how a given solution might perform. As stated before, more granular level type records in theory yield better performing data mining solutions.

So far, we have discussed data sources that are available at an aggregate level for targeting prospects for acquisition programs. Individual-level data for acquisition programs can also be purchased. One organization has built a massive consumer database of individual-level data through online surveys to Canadians. The incentives for consumers to fill out this survey are coupons and discount offers for many different products. One argument, though, against using this data is that the information is self-reported and that what people say does not necessarily reflect what they will do. Another argument against using this data is that there may be a responder bias. In other words, people who complete the survey may not be representative of the population comprising our initial audience. Yet, even with these limitations, this information has been of great value for rank-ordering and differentiating prospects and ultimately better targeting of prospects for acquisition programs. Marketers can use this information in two ways; they can rent names based on this information, or they can build models based on the information to then identify the best list of prospects from this database. Whether to rent or to model the names will depend on the type of business and a company's products.

In business-to-business marketing, the number of vendors offering data enhancement services for better targeting is growing. Some companies offer these services in particular sectors while others deal with large companies only. But there are two main players that offer data enhancement as well as list rental services for all companies regardless of industry sector or size. This information consists of individual company records and includes details on industry sector, industry sales, number of employees, years in business, and a range of other firmographic information. Marketers again have the choice of renting names specifically for a particular initiative or of using the firmographic information to model these names for that same initiative. Once again, whether to rent or model names will depend on the type of business and a company's products.

MORE ABOUT MISSING VALUES

The transformation of missing values, as noted earlier, is a critical process in data mining. Inputting an average for missing values of continuous variables or using a default value for missing values of categorical variables is acceptable in many situations. But a more exhaustive and robust approach might be considered for variables that are known to be highly significant in a typical data mining process. For example, certain insurance products have higher appeal depending on the age of the buyer. In some cases, age might be the strongest predictor of response to this insurance product. However, we may only have age information on 50% of the individuals in the database. Inputting an average for the remaining 50% would reduce the variable's ultimate impact on any targeted behavior as we are diminishing the impact of any trend based on the continuous nature of the variable. Using predictive modeling techniques or some type of CHAID analysis, we can impute or estimate the age of the records where this value is missing based on other information. Although age is not a "hard" observed age but only an estimate, it is still better to have estimates that more closely resemble the continuous nature of the variable rather than impute one value, such as an average, to the entire group of records that have missing values in that field.

CHAPTER 11

SEGMENTATION

TO UNDERSTAND CUSTOMERS, THE FIRST STEP SHOULD BE TO SEGMENT customers. Although consultants agree on this first step, they do differ on the approach to segmentation. There are practical approaches and also scientific ones to segmentation. Rather than simply choosing to adopt one approach over the other, the correct approach will depend on the complexity of the given customer base and on the requirements of the current business challenge. For instance, a bank's customer database containing 1 million customer records may require a much more sophisticated approach than a retail customer database containing 50000 records. Identifying students as a customer segment for a student program in a bank is going to be far less complex than trying to identify unique customer segments where the objective is to create distinct corporate strategies for these segments.

When adopting an approach to segmentation, it is important to remember the old business adage: KISS, keep it simple, stupid. This has tremendous relevance for the following reasons:

- Easier to understand
- Easier to execute
- Easier to track

In developing any segmentation system, the incremental benefits of the system must be weighed against its complexity. Complexity makes it more difficult to understand what the segmentation means throughout the organization thereby making it more difficult to execute the segmentation system and track the results.

The first consideration should be the number of segments to be created in a database. Some organizations have their customer database segmented into over 50 segments. In adopting or developing an initial segmentation system, a good rule of thumb is plan for no more than 10 segments. In keeping the number of segments to a minimum, initial learning can be acquired that may provide insights supporting either an increase or a decrease in the number of segments.

Some critics may argue that this is too broad in terms of identifying the right number of customers for a particular program. However, segmentation systems are designed to develop a number of broad segments with specific models that could be applied to each system.

Segmentation can mean different things; for example, it may mean simply using business rules, such as select all customers who have been a customer for more than 2 years, have incomes of more than $100K, and have bought more than two products. Or segmentation can mean scoring customers through a model or perhaps a value metric and selecting customers based on score. Another scenario may involve the use of statistics to determine homogenous groups of customers. No one approach is necessarily superior to any other. They are all good approaches; which one is more appropriate depends on the particular business problem and, more important, on the data. Simple and complex solutions can be equally acceptable to a particular business stakeholder. Not all segmentation requires sophisticated statistics. Sometimes, a simple and pragmatic approach will suffice.

SIMPLE VERSUS COMPLEX APPROACHES

The first consideration in choosing between simple and the complex approaches is the data environment. More data implies a higher quality of analysis and more purposeful application to marketing activities. This in turn will depend on the quality of the data itself as well as on its variety. The more complex solutions will arise in data-rich environments.

The second consideration is the volume of customers. A larger number or volume of customers offers more potential for deriving greater benefits even when lift becomes more marginal. For example, comparing a 1% business lift on 2 million customers to a 5% lift on 200 thousand customers shows that the larger opportunity is with the larger volume of customers despite the lower lift potential. With this larger volume of customers, more complex solutions can be explored.

Let's take a look at an example to bring some of these scenarios to life. Suppose a company sells one product and collects customers' billing information for the past two years as well as name and address. The current number of customers is 100 thousand. The company must determine how this information is to be used, specifically, whether it will be used for targeting or merely for communication purposes? Most people would say both, but in reality, a specific solution will tend to focus either on targeting or communication but not equally on both objectives. For instance, if targeting is the prime objective, clustering is not required. Instead, names could merely be rank-ordered based on the desired business objective, in this case billing amount or sales. Rank-ordering places customers into groups or deciles; the top decile represents the highest billed customers, and the bottom decile represents the lowest billed customers. Groups of customers can then be selected for a marketing initiative based on their prior total sales (billing) with the company. However, if marketers want to establish more meaningful communication with this company's customers, they would ask for other information besides sales to help in this process. In the example, there is no opportunity to use other information for communication purposes since no other information beyond billing amount exists.

However, if this same company collects payment type and keeps track of a given customer's billing history, then it is possible to create two other key variables. The first one, tenure, is developed using the date of the customer's first billing as a proxy for tenure. Meanwhile, payment type can be used to separate customers into preauthorized payments versus those are not preauthorized. Other information may be gathered, and the target group of customers may be identified as the top 50% based on total customer billings in the past year. Marketers then must decide whether this tenure and payment type information is rich enough to develop unique communication programs. Indeed, a simple clustering exercise might help to decide this in a quantitative manner by scientifically determining whether there are some distinct customer groups based on tenure and preauthorization plans among these top billers. The data and results might look as shown in figure 11.1.

	Tenure Average	% with pre-authorization plan
Cluster 1	2.5 years	40%
Cluster 2	6 years	20%

Figure 11.1 Example of 2 Cluster Solution on Top 50% of Customer Billers

In this simple example, the clustering exercise provides additional insights. Here marketers can develop a communication program to newer customers who are more likely to pay by automatic withdrawal versus a communication program to customers with longer tenure who pay by check or credit card.

LOOKING AT THE COMPLEX EXAMPLE

In the above example, it can be determined whether tenure or preauthorization plan has an impact on being a top biller. Through profiling, we compare variables such as tenure and preauthorization against the top 50% billers as opposed to the bottom 50% billers. If either or both of these characteristics have an impact on being a top biller, then communication programs could be designed for the top 50% without doing any clustering. Yet, if profiling is the more reasonable and practical option in this case, all customers could be segmented based on value; and marketers could then simply distinguish the high-value customers from regular customers rather than develop clusters around the high-value group. In many cases this is an acceptable solution. However, for companies with large customer volumes (i.e., much larger than 100000 customers) and very rich data, this may not be the optimal solution. For example, a bank may have 5 million customers with the top 2.5 million customers deemed high enough in value to be considered the net eligible group for future CRM initiatives.

Profiling these top 2.5 million customers compared to the bottom 2.5 million reveals six key characteristics that best differentiate the two groups:

- High tenure
- Live in Toronto
- Have a mortgage
- Have several investments
- Have several loans
- Have a Visa card
- Have high transaction fees

Analysts could argue that this is a profile of a high-value customer that could have been created without any data mining. The question of how to use this information to develop unique communication programs remains. How do marketers approach this target group of 2.5 million customers?

	Cluster 1 Average	Cluster 2 Average	Cluster 3 Average
High Tenure	3	10	3.5
Live in Toronto	0.8	0.5	0.45
Have a Mortgage	0.9	0.4	0.49
Have Multi Investments	0.3	0.35	0.7
Have Multi Loans	0.4	0.5	0.8
Have Visa	0.6	0.9	0.65
Have High Transaction Fees	0.5	0.8	0.55

Figure 11.2 Example of 3 Cluster Solution for Bank

This is where clustering plays a significant role. For example, the analysis could provide the types of clusters shown in figure 11.2.

This information shows that three separate programs could be developed to high-value customers. By comparing the variable means between clusters, we can identify certain variables for a given cluster where their variable means are different from those of the other clusters. For cluster 1, programs would be developed that communicate the advantages of different types of long-term lending schemes (mortgages) as well as particular financial benefits that are more advantageous to someone living in a large city like Toronto. For cluster 2, programs might be developed that speak to longer-tenured customers about benefits to credit card users as well as benefits to high transaction users. Meanwhile, cluster 3 programs could be developed for the type of customer who is more financially sophisticated both in terms of investments as well as loans.

This type of approach allows marketers to use the information in a two-pronged manner to target the best customers based on a defined metric and also use other information to meaningfully communicate with them.

Profiling versus Clustering

As discussed above, in addition to clustering, another option might be to differentiate the top billers (top 50%) from the bottom billers (lower 50%). The difference between profiling and clustering is that profiling uses an objective function, namely, whether or not the customer is in the top 50% of billers. Other customer characteristics are then analyzed to identify those that are key characteristics in differentiating a top-billing customer from the bottom 50% of customers. This approach is called supervised learning because the objective function of being in the top 50% or not determines (supervises) what these key characteristics are that differentiate the top 50% of billers from the bottom 50% of billers.

In clustering, there is no one variable determining the outcome of the other variables. The objective of clustering is to identify those characteristics that best assign customers into unique and distinct groups. The objective here is not to optimize a specific variable or metric but rather to maximize variation between customer groups or clusters and minimize variation within customer groups or clusters. The data or variables in this case are unsupervised or not analyzed against a specific customer variable or metric.

That is, cluster analysis is a type of analysis considered unsupervised learning, and any learning or insight from this type of analysis is not assessed against a certain piece of information. As discussed earlier, response models are considered supervised learning because the analysis is aimed at the objective of optimizing response. Retention models attempt to find key triggers or variables geared to predicting retention. With cluster analysis, the learning is unsupervised in the sense that the analysis aims to uncover trends or patterns from all data with the objective of classifying individuals into distinct homogenous groups. The key difference to supervised learning is that the classification in cluster analysis is not determined or based on any one specific field but on all the data fields.

What does cluster analysis actually do? In practical terms, cluster analysis attempts to classify records, usually individuals for the purposes of CRM marketing, into distinct groups where each group is homogenous or similar in terms of characteristics. In other words, minimal variation of characteristics or behavior exists between individuals within the groups but variation of characteristics or behavior between the cluster groups of individuals is maximized. This implies that the characteristics of one given cluster are indeed very different from those of the other clusters, as for example in figure 11.3.

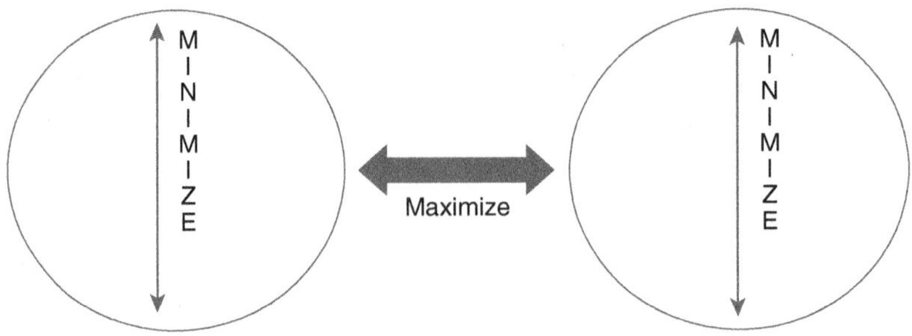

Figure 11.3 Schematic of How Cluster Variation Works

Cluster analysis does not consist of just one technique, but many. Ongoing research in statistics continues to explore the validation and value of new approaches. Two of the more common techniques are k-means clustering and hierarchical clustering. For these two approaches, it is imperative that the data be standardized. Without standardization, two conditions arise that can affect results in a negative manner. First, outliers in the data can skew results by extreme values of a given variable. Clusters are created based on variance of the data, and this variance is determined or based on means or averages of all the variables in the data. If some of these variable means or averages are distorted by outliers, it would make sense that these distorted variables could yield misleading results in any cluster solution.

Second, the magnitude and scale of variable values can affect results as can the fact that the magnitude of variables' values will vary dramatically in any cluster solution. Two common variables that might be part of a cluster solution are income and age. Income can range from 0 to millions of dollars while age typically ranges from 0 to 100. Again, cluster analysis is sensitive to actual values, and it is evident that the average or mean and resulting variance values would be very different for income and age. But these differences would be based on the magnitude or scale of the data rather than on how each field varies relative to the other. Obviously, there must be a way to look at how data varies in a relative manner, thus allowing a comparison of variables that is like comparing apples to apples rather than like comparing apples to oranges.

Both these situations of outlier values and the magnitude of values must be adjusted prior to any cluster solution. The solution is to standardize the data prior to conducting any clustering exercise. There are a number of ways to standardize the data. The most common approach is to normalize the data through the use of z-score. In this case, the normalization of data means calculating the actual z-score statistic variable value in the customer record. The actual numbers in effect range from -3 to +3 with larger absolute numbers more statistically significant (+ or – 1.96 = 95% confidence interval and + or – 2.58 = 99% confidence interval). The critical formula for calculating the z statistic is: (Actual Value-Mean Value)/ Standard Deviation. This represents the foundation of most business applications in assessing the significance of a given business situation.

This formula essentially calculates the number of standard deviations. It is this standard deviation unit that can be compared across all variables. Variables that have tremendously differing scales, such as age and income,

Age	18 19 20 21 100
Rescaled Age	-1 -.99 -.98 . . . +.97 +.98 +1.

Figure 11.4 Scaling Age Values to Range between -1 and +1

can now be compared or used in any analytical solution due to the standardization of values through this approach.

The second approach to normalization is less mathematical in the sense that each variable with its range of values is rescaled to a range between -1 and +1. An example of what this means for age values is shown in figure 11.4.

For age 25, algebraic calculation yields a normalized value of -.83 while for age 75, the normalized value is .75. This second approach is somewhat more limited in the sense that the distribution of records across this range of values is completely ignored. However, both approaches provide options when it comes to normalizing the data, and the preferred option is the use of the z-score statistic.

Let's now look at the concept of clustering in more detail. In clustering, centroids are produced that represent the center point of the cluster. The cluster and its centroid put records where the difference between the actual value of a given variable and the mean value of that variable are minimized within that cluster, and the mean values of that particular variable between clusters are maximized. The centroid consists of a multidimensional point that represents the means and averages of all variables in a given cluster. These cluster routines are iterative for a given cluster and its centroid. The iterative nature of the routine finds observations with values that are closest to the centroid within its own cluster (homogeneity) while attempting to maximize the difference between the observation's values from the other cluster's centroid values (heterogeneity). In k-means clustering, the number of initial clusters represents the number of records. The routines work backward by attempting to reduce the clusters to a number where again the variable values within clusters are minimized while variable values between clusters are maximized.

Hierarchical clustering works in an opposite fashion where the initial set of clusters is one. It then moves forward by attempting to find the optimum number of clusters based on how variation is calculated as indicated in the previous paragraph.

But the question remains as to what this optimum number is. This can be rather subjective given the underlying objectives of cluster analysis. For the perfect solution the number of clusters would equal the number of records. Obviously, no real insight or intelligence is provided here. There must be a balance so that the ideal number of clusters provides meaningful insight and intelligence while achieving the cluster routine's objectives of maximal variation of information between clusters and minimal variation of information within the clusters. One way to look at this is by exploring the R^2 or explained variation (cubic clustering criterion). As the number of clusters is increased, the percentage of explained variation will continue to increase until it is fully maximized when the number of clusters equals the number of records. But as stated before, this is impractical, and no insight or intelligence is gained with this so-called perfect solution. The challenge then is to find the optimum balance in terms of maximizing explained variation and providing the optimum level of insight and intelligence. One approach, called the elbow approach, is to plot the level of explained variation across each cluster solution, as shown in figure 11.5.

As the number of clusters increases, the level of explained variation increases. From the above table, each cluster solution is plotted against the level of explained variation. In terms of explaining variation, this curve begins to flatten out. The point where the curve begins to flatten out is the inflection point, the point where the number of clusters reaches the ideal solution. In the case above, this ideal number may be 3 since the percentage of explained variation begins to flatten out or, in mathematical terms, the actual marginal increase in explained variation begins to decrease.

In practical terms, analysts might focus on three clusters as being one option, but other options, such as solutions with two or four clusters could also be explored. These other options would also be in close proximity to the derived inflection point of the three-cluster solution. This is where the art of data mining is evident as analysts explore the means of all the variables

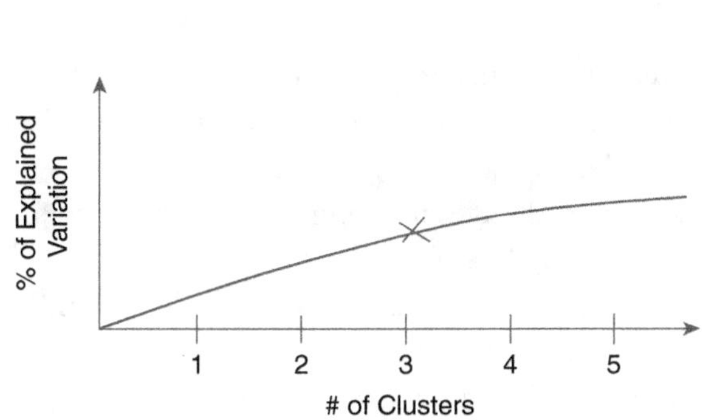

Figure 11.5 Using the Elbow Technique to Define the Optimum Number of Clusters

within each cluster solution as described earlier in this chapter. Based on the analysts' and business user's judgment, a solution is then arrived at that makes the most practical sense from a business perspective, a solution that would not deviate too far from the preferred scientific answer.

Rather than delve more deeply into these techniques and the mathematics, it is more important to understand the overall mechanics of clustering and how it can create solutions for a given business initiative. Virtually all data mining software packages contain clustering routines. In SAS, one routine that quickly provides a solution is appropriately named FASTCLUS. Here users can quickly look at a variety of different cluster solutions to find the inflection point where the marginal increase in variance is maximized.

Once solutions are produced, the next challenge is the communication aspect. In other words, what does this solution mean to the business? An example of a clustering exercise that was done for customers of a wealth management company is shown in figure 11.6.

In this example, the final solution was a six-cluster solution or a system that grouped customers into six segments. We created descriptions for each customer segment by identifying those variables within the cluster that were statistically different from variables in the other clusters. We considered over 200 variables in this exercise. In order to support our description, we reported the variable means of the statistically significant variables within the given cluster and then compared them to the variable means of the other clusters. This was done for all six clusters. In the table in figure 11.6, only the result of the cluster 1 solution and those variables that were statistically significant are shown. The mean of each of these variables is very different from the variable means in all the other clusters. This indicates that these variables can be considered unique for this cluster. These variables can then be used to describe the characteristics of the cluster. In this example, the results of cluster 1 provide the following profile:

- Do not report their risk tolerance level (MISSRISK)
- Do not like to report their occupation on the KYC form (MISSEMPLOY)
- Have not had a change in a long time in any of their engagements with the company (LSTCHGYR)
- Put more money in Canadian growth funds (CGFPP)
- Like to reinvest (REINVEST)
- Put more in Canadian growth income funds with income (CGIFPP)
- Are less likely to contribute to an RRSP (CONT2002)

Cluster 1 Variables that are significant								
Variable	**Variable Description**	**Cluster 1 Mean**	**Cluster 2 Mean**	**Cluster 3 Mean**	**Cluster 4 Mean**	**Cluster 5 Mean**	**Cluster 6 Mean**	**Correlation Coefficient**
MISSRISK	RISK TOLERANT	**0.97**	0.54	0.48	0.03	0.59	0.52	0.7303
MISSEMPLOY	OCCUPATION CODE IS MISSING	**0.27**	0.1	0.08	0.07	0.04	0	0.2735
LSTCHGYR	# OF YRS SINCE LAST CHANGE	**4.11**	2.86	2.25	2.83	3.81	3.16	0.2478
CGFpp	% IN CANADIAN GROWTH FUNDS	**0.46**	0.36	0.36	0.26	0.4	0.36	0.1968
REINVEST	REINVEST	**0.02**	0.01	0.01	0	0	0	0.0735
GGIFpp	% IN CANADIAN GROWTH INCOME FUNDS	**0.02**	0.01	0.01	0.02	0.02	0.01	0.0485
CONT2002	$ CONTRIBUTION IN 2002	**58.06**	355.23	319.1	203.54	203.54	135.64	-0.0621

Figure 11.6 Wealth Management Company Cluster 1 Results

This same exercise is then conducted for all of the clusters with the end result being a profile description of what these records look like given that they reside in a certain cluster. In this cluster (cluster 1), the description might be that this segment is comprised of customers who are less engaged with the company but more likely to reinvest and in growth funds but not in RRSPs.

Clustering can provide tremendous insight regarding the characteristics of a homogenous group of customers. However, business judgment is even more important in this exercise than in other data mining exercises because the description of the cluster is based on how a business user interprets results. There is some subjectivity in this. In a typical situation, analysts and marketers interpret the results and use their judgment to describe the cluster based on comparing the averages and/or median values across the different clusters. There is much more subjectivity in clustering when compared to predictive modeling. This is why cluster descriptions can be useful but in many cases do not reveal the complete picture of the cluster. In many cases, particularly in a solution with many clusters, descriptions used in one cluster will often overlap with those of other clusters. This provides a clear rationale for considering cluster solutions as a broad segmentation solution.

CHAPTER 12

APPLYING DATA MINING TECHNIQUES

THE THIRD STAGE IS WHERE THE ANALYTICS (BOTH ADVANCED AND non-advanced) occur. Segmentation, which was discussed in the previous chapter, also occurs in this phase. Predictive analytics' solutions as well as business and campaign analysis all occur in this phase. Arguably, this is where the "exciting" part of data mining happens. We have moved beyond the data phase, which in practical terms is more critical than this phase in achieving project success. Data is the key ingredient, but in this phase analysts apply tools in order to bring the "magic" out of the data. We will commence this section by discussing a number of techniques.

REGRESSION ANALYSIS

One of the really fascinating diagnostics that arises from regression analysis is output relating to the power of a certain variable in an overall predictive model. In the typical OLS (Ordinary Least Squares) regression technique, the diagnostic is often referred to as the partial R^2. Total R^2 represents the power of the model with 1 representing a perfect model where all the variability of the target variable is completely explained by all the variables input into the model. There is no random error. Meanwhile, a value of 0 implies that the model is completely ineffective in explaining the variation of the target variable. The variation of values of the target variable is all due to random error. Total R^2 essentially represents the explained variation accounted for by the model inputs divided by the total variation (explained variation + random variation). A partial R^2 represents the variable's contribution to the explained variance of the entire model.

By dividing the partial R^2 into the total R^2, we can determine a diagnostic that is the variable's contribution to the overall model. These diagnostics can convey very interesting insights to marketers; an example of a response rate model is shown in figure 12.1.

In this example, marketers are trying to sell a premium credit card to banking customers. The behavior score has the most impact (35%) in the model. Other variables related to customer engagement (i.e., banking activities); all exhibit a propensity by the customer toward buying a premium credit card. The notable exception to this is having an RRSP, which may indicate a mind-set that customers are "borrowers" or "seekers of credit" rather than "savers." This is only an opinion, but it represents a notion that could be tested in a future campaign. The other variable (are female) indicates that females are less likely to purchase this premium credit card.

Another example is shown in figure 12.2; here, one variable (have an RRSP accounts for 80% of the model's power. The variable is also binary indicating that there are only two outcomes (1: have an RRSP or 0: not have an RRSP). This model has essentially only one variable. But is this

Model Variable	Impact on Response	Contribution to Overall Equation
Behavior Score	Positive	35.00%
Average Score	Positive	25.00%
Have an RRSP Product	Negative	15.00%
# of Fin. Inst. Products	Positive	10.00%
Avg. % of Credit Limit Used	Positive	10.00%
Live in Prairie Provinces	Negative	5.00%

Figure 12.1 Contribution of Variable to Overall Model

Model Variable	Impact on Response	Contribution to Overall Equation
live in Quebec	positive	80.00%
tenure	negative	8.00%
Recency of last purchase	negative	7.00%
# of products purchased	positive	5.00%

Figure 12.2 Example of Model's Power Explained Predominantly by One Variable

really optimal? A more reasonable next step would be to create two up-front segments:

- Have an RRSP
- Not having an RRSP

The analyst would then develop separate models for each segment. In fact, this is a way of integrating segmentation and modeling in order to create better targeted solutions. However, it was the use of R^2 statistics that provided the learning in determining these up-front segments.

In producing models, being able to demonstrate the relative power and impact of each variable in a solution allows marketers to get inside the black box of modeling. "Black box" here means that it is not known how the key components of a given solution are determined. It is extremely important that data miners as well as marketers and business analysts understand the key components. Relying solely on the solution's ability to deliver results without some basic understanding of its components is a recipe for disaster. Furthermore, the ability to empower more people with the information on what comprises the model widens the knowledge base for assessing the given solution. Based on this wide array of experience, data miners may find that a current model variable should not be used. For example, the purchase of product X may be discovered as a model variable with a positive impact on response to product Y. Upon further investigation, analysts discover that for the past two years, product X has been given as a free incentive to new customers, and this incentive is now going to be eliminated. This information provided by the business person or marketer would suggest that this variable should be eliminated in predicting response. As new customers seem to respond more, it would be very likely that a variable such as tenure should be a replacement; if tenure is already in the current model, its overall power would be strengthened. These situations illustrate the point that this type of information should be shared with other business stakeholders in the project. Acknowledging that modeling is black-box technology with no understanding of the solution's key components is unacceptable in today's business environment where knowledge is power. Organizations that can effectively communicate the technical details of a given solution to broader groups of stakeholders within the organization will develop solutions that are built by the team and not in isolation by data miners or the marketing department.

FACTOR ANALYSIS

Factor analysis is a commonly used statistical technique for data reduction. In business applications analysts may have hundreds of variables. Through the use of factor analysis, analysts can reduce the number of variables by using factor output scores as potential variables. For instance, one project may involve 200 variables that are now accounted for by 17 factors based on the eigenvalue cutoff (the eigenvalue definition will be discussed later on in this chapter). These 17 factor scores can then be used as input variables for an analysis or modeling exercise. Because of the difficulty in explaining the output in meaningful terms to business stakeholders, the option of using factor scores as model variables is not used. However, the more common use of factor analysis is to select variables with the highest factor loading scores in a given factor. An example of this is shown in figure 12.3.

In this simplistic example, there are three factors that best explain the nine variables. In this example, the variables (income, education, and wealth) in factor 1 are the selected variables to represent this factor. Factor 1 might easily be described as representing a measure of affluence. In the second factor, it would appear that three products (A, B, and C) have the strongest factor loading coefficients. Upon further investigation, analysts may discover that these three variables essentially comprise a common product category of men's shoes. With this knowledge, a binary (yes/no) variable of men's shoes could be created as the variable best representing the second factor. The third factor reveals that once again three products (D, E, and F) have the largest factor loading coefficients. These factor

Variable	Factor 1	Factor 2	Factor 3
1) Income	0.905	0.255	0.255
2) Education	0.855	0.373	0.212
3) Wealth	0.956	0.303	0.185
4) Product A	0.303	0.855	0.205
5) Product B	0.295	0.805	0.245
6) Product C	0.323	0.755	0.285
7) Product D	0.105	0.355	0.755
8) Product E	0.155	0.405	0.705
9) Product F	0.085	0.304	0.725

Figure 12.3 Factor Analysis Output

coefficients represent the product category of sporting goods. Using this knowledge, another binary variable (yes/no) could be created comprising the product category of sporting goods.

Factor analysis can be extremely useful as an analytical tool. Users can vary the number of factors generated by using different eigenvalue cutoffs. This flexibility allows users to examine what information and insights are obtained with few factors as opposed a large number of factors. As the number of factors increases, variables and information are observed that are most important in explaining why factors repeat themselves since they have already been used in explaining a prior factor. In examining the different factor number options, users can then ascertain how much additional information is gained when factors are added or, conversely, how much information is lost when the number of factors is reduced. As we have seen with much of the output and results from statistical analysis, analysts use both statistics and their own insights and domain knowledge about the business to derive the optimal answer. The optimal answer is not totally driven by statistics nor should it be.

In defining the number of factors, users utilize the eigenvalue as a key statistical threshold in determining the number of factors. In most statistical software programs, this threshold is 1. If the value of 0 is substituted for the eigenvalue, the number of factors equals the number of variables input into the factor analysis routine. This is expected since an eigenvalue of 0 would imply that the routine is not utilizing any mathematics or statistics to perform any data reduction. As stated earlier, in any data reduction routine better insights and information regarding the analysis is gained, but at the same time information is lost. The eigenvalues enable users to determine how much total information is lost by using fewer factors. As stated before, the eigenvalue is a statistical diagnostic that captures the total information explained by the factors. More factors result in lower eigenvalue cutoff scores while the opposite is true with fewer factors. The total eigenvalue could be plotted against each factor number option.

Just as the optimum number of clusters had to be determined, analysts must find the inflection point where the increase in loss of information becomes marginal, and use this as the optimum number of factors. Adding in incremental factors does not produce a noticeable decrease in the eigenvalue. In other words, the marginal impact of adding more factors is not really providing information gain. The raw statistical output will be used in conjunction with analysts' own insights and experiences based on fewer factors as opposed to more factors. These insights and experiences may lead analysts to select the optimum number lower or higher than what science suggests. But analysts' selected number of factors will not deviate too much from the optimum number of factors determined based on the graph in figure 12.3 because the optimal data mining answer is a combination of science and practical insights. In most situations, analysts will collaborate with business persons regarding insights from the results.

Users can apply these techniques in predictive modeling scenarios that offer a significant opportunity to reduce the number of variables. The same scenario would also apply to clustering where users may have to deal with hundreds of input variables. In fact, factor analysis is a standard procedure in clustering because clustering routines become less robust as more variables are introduced.

Routines such as factor analysis can be used to determine the reduced set of variables (40–50) that is the typical range of variable inputs into any clustering exercise. In predictive modeling, factor analysis is an option but not a requirement for building the solution. The most commonly used technique in factor analysis is called principal component analysis, commonly referred to by its acronym PCA. There are other approaches to factor analysis, but they all have the common objective of reducing data and achieving this through unsupervised learning, which has been discussed in the previous chapter. In other words, the technique maximizes the variation of the data set by organizing the data into patterns and not against observed behavior. Because this variation is not maximized against an observed behavior as in a predictive response model, the learning or derivation of the factors is unsupervised. Each pattern or factor is completely independent from the others. No multicollinearity (i.e., no correlation) exists between factors. In determining the number of factors produced through the routine or algorithm, users have control of this through the values of the statistical criteria they input into the routine, such as the eigenvalue metric discussed above.

Debates about these routines continue among academics, many of whom maintain that factor analysis is limited by its assumption of linearity. Most distributions of data are not purely linear. Ongoing research is attempting to produce data reduction routines that are nonlinear in nature. From a purely mathematical or academic standpoint, the nonlinear routines will be superior to the linear ones. From a business standpoint, though, the real advantage will be the incremental value that this provides to a given data mining problem as compared to what traditional factor analysis conducted on the same data mining problem offered. As always, this incremental advantage in the business world will be measured by the additional dollar benefits due to the new technique.

DEVELOPING THE SOLUTION

As discussed in previous chapters, creating information that can be used in any data mining solution is critical. The derivation of new fields and variables represents a key area of data miners' insight and expertise in the overall project. In fact, the creation of this data environment represents by far the most time-consuming part of the exercise. This makes sense as it is in this stage that data miners will fundamentally determine the quality of a given solution. The use of a variety of different statistical techniques will yield more or less the same performance. Yet, in most practical situations, it is the data and not the mathematics or statistics that determine the quality of a given solution.

Data and the creation of the analytical file are important, but something else must be done with the data. In other words, how do data miners work the data? Do they want to produce reports or conduct some ad hoc analysis or engage in more rigorous forms of mathematics in terms of developing more sophisticated solutions, such as models, profiles, or cluster segments?

Before conducting any analysis or developing more sophisticated solutions, the goal of the data mining project must be identified. For instance, is the analysis intended to gain insights and learning about a given customer behavior? If so, the analysis is directed or supervised against this behavior and is called supervised data mining. When the analysis is directed or supervised against a given behavior, data miners call this behavior the dependent variable or objective function. Other situations require analytics but the analysis is not directed against a given objective function or dependent variable. Some examples of this are cluster analysis and factor analysis, discussed above. Both technologies (cluster and factor analysis) can analyze reams of data (often hundreds of variables at a time), but both analyses are not directed by any specific customer behavior. In more practical terms, if data miners are asked to conduct a customer attrition analysis, then the data mining is directed because they will identify a specific field that relates to attrition such as "cancel reason." They could then potentially analyze all other customer-related fields against the above-mentioned field. This approach is also similar to predictive models. Predictive models require that users identify what the model is trying to predict. For example, a predictive response model would require analysts to create a field in the analytical file that indicates response (1: yes, 0: no). This response field becomes the dependent variable or objective function of response. In simpler terms, all the analyses in building the model are directed at trying to find information that best predicts the objective function of response.

Having determined what analysis is required, data miners then use a variety of tools available in this stage of the analysis. The choice of tool depends on the specific business need. For instance, analysts often simply have to produce reports. No sophisticated statistical algorithms are required; the technical need is simply to manipulate and organize the information in order to produce the required reports. In building a predictive model, though, a variety of statistical tools can be used, and the selection of a given tool will depend on the specific task data miners are to perform. For example, in building a model, analysts must first get a sense of what variables and information most impact the desired modeled behavior.

Correlation routines serve as the first statistical tool that can yield these initial insights. These correlation reports, though, do not represent the typical correlation matrix reports most statisticians are familiar with. Traditionally, conducting a correlation analysis on five variables leads to a 5x5 matrix, as shown in figure 12.4.

In this correlation matrix we observe the relationship (weak and strong) between all possible variable pairs. The strength of the relationship is determined by the absolute value of the correlation coefficient as depicted in each cell of the matrix. The sign of the coefficient indicates simply whether the trend between the two variables is negative or positive. In the example above, the strongest relationship occurs between spending and the number of products with a correlation coefficient of .7. As

	Live in Quebec	**Spend**	**Number of products**	**Recency of Spend**	**Credit Score**
Live in Quebec	1	-0.5	-0.65	0.3	0.4
Spend	-0.5	1	0.7	-0.45	0.5
Number of products	-0.65	0.7	1	-0.52	0.55
Recency of Spend	0.3	-0.45	-0.52	1	-0.24
Credit Score	0.4	0.5	0.55	-0.24	1

Figure 12.4 5 × 5 Correlation Matrix

spending increases, the number of products bought increases. Meanwhile, the strongest negative relationship (-.65) is between living in Quebec and the number of products; this indicates that people living in Quebec overall bought fewer products. In this matrix, correlation coefficients of 1 are shown for those variables correlated against themselves.

However, in data mining correlation analysis is not shown in a complete matrix as above, but in a report analyzing all variables against the dependent variable or objective function.

For example, if we had seven variables that included response behavior as the behavior to be optimized, then the correlation report would consist of a 6 x 1 report. Here the variable to be predicted (response) is listed on the horizontal axis and all other variables (independent) are listed on the y-axis. In this report, all variables on the y-axis are ranked by the absolute value of the correlation coefficient against the variable listed horizontally on the x-axis. As stated above, negative values are as good as positive values. Besides the correlation coefficient used to rank the other variables against the variable listed horizontally, there is another column called the confidence interval. This confidence interval is a metric relating to the statistical significance of the variables on the y-axis as opposed to the variable on the x-axis or our dependent variable. Basically, the larger the correlation coefficient, the larger the confidence interval is. The sign of the variable indicates whether there is a positive or negative impact with the x-axis or dependent variable. Figure 12.5 shows a correlation matrix used to develop a predictive response model. The x-axis variable is response and does not need to be shown since the table implies that the correlation results are measured against response.

In this simplified example, the report indicates that age, tenure, number of products purchased, and number of promotions since last purchase are statistically significant against response. Meanwhile, household size

Variables	Correlation Coefficient	Confidence Interval
Age	-0.0673	0.995
Tenure	0.055	0.98
# of products purchased	0.045	0.97
# of promotions since last purchase	-0.031	0.96
Income	-0.0045	0.5
Household Size	0.001	0.2

Figure 12.5 Correlation Matrix of 6 Variables versus Response

and income are not statistically significant against response and would not be relevant in a modeling solution. In this example, analysts are building a response model. The results of the correlation analysis indicate that the statistically significant variables should be considered in any data mining solution. Clearly, any data mining solution would explore the notion that younger people with longer tenure and who have bought several products but who have received fewer promotions are more likely to respond. With correlation analysis, analysts essentially have an initial mathematical glimpse of the data. The mathematics provides the quantitative support for formulating a data mining solution.

However, the analysis is far from complete at this stage. Further manipulation of variables can occur based on the insights of the correlation analysis. Let's suppose that one variable, such as age, had a correlation coefficient of .3 while all the remaining statistically significant variables had correlation coefficient values of .10 or less. Due to the scale of difference between age and all the other variables, this information might tell us that segmentation based on age should occur prior to the development of any models. CHAID (Chi-Square Activation Interaction Detection, which will be explained more fully later on is often used as a decision-tree tool in building models. As the tool can be used to define the actual variable breaks within a branch level, it can also be used to define the optimum breaks for a given variable. In our example above, CHAID would be used to determine the optimum age breaks that would then be used as our segment definitions, as shown in figure 12.6.

For each segment, separate models could be developed. Yet, this insight and the use of CHAID might never have been considered if the correlation analysis had not revealed the age results.

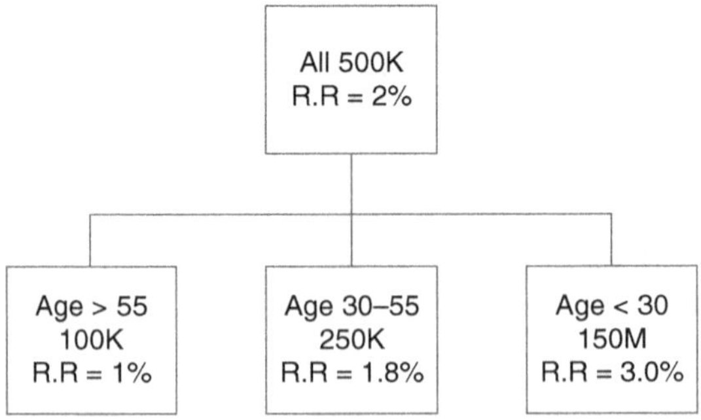

Figure 12.6 Example of CHAID Used to Define Model Segments

Beyond correlation, one might consider CHAID another tool in deriving new variables. What is meant by this? As previously mentioned, variables with many different character value outcomes (otherwise referred to as data granularity in the negative sense) are not useful in data mining. An example is product-level information where there could be hundreds of different products that are sold. In data mining, this would mean creating hundreds of binary variables (1 if a product was purchased and 0 if a product was not purchased). Many of these newly created binary variables would be meaningless in any data mining exercise. The number of occurrences within that variable (the 1s in this case representing customers who purchased that specific product) would be very sparse and very likely insignificant in any data mining exercise. Let's take a look at what exactly this means.

As a first step analysts would typically create binary variables for each potential outcome. If there were 200 product categories, 200 binary product variables with yes/no (1/0) outcomes would be created. Now suppose the company has 200,000 customers. Calculating an average product purchase using just the above information, we might estimate that .5% of the customers representing on average 1,000 individuals actually purchased a product from the list of 200. This .5% is simply too small to identify any meaningful trend or pattern.

But what if there was a way to group character values together? Suppose in the case of a retailer, products could be grouped together into some broader category. By grouping these occurrences together, in essence there are more 1s (occurrences of purchase within the broader product category) or critical mass around a given occurrence.

CHAID provides the key tool for grouping these outcomes in an optimum manner. The grouping of these outcomes would be based on the behavior to be optimized. In the example above, this would be response. An example of a CHAID analysis of product categories is shown in figure 12.7.

For the sake of simplicity, only 10 product codes are used in this example, and they are represented by letters (A–J). In this example, the product codes (which are in letters) are grouped into optimal broader categories or nodes by the CHAID routine when analyzed against response. With this information, binary variables (0, 1) could be created for each of the three nodes. But a superior approach, though, would be to create a product index variable with three outcomes:

- 1: For the node representing A, D, B, and C product categories
- 2: For the node representing H, E, I, and J product categories
- 3. For the node representing G and F product categories

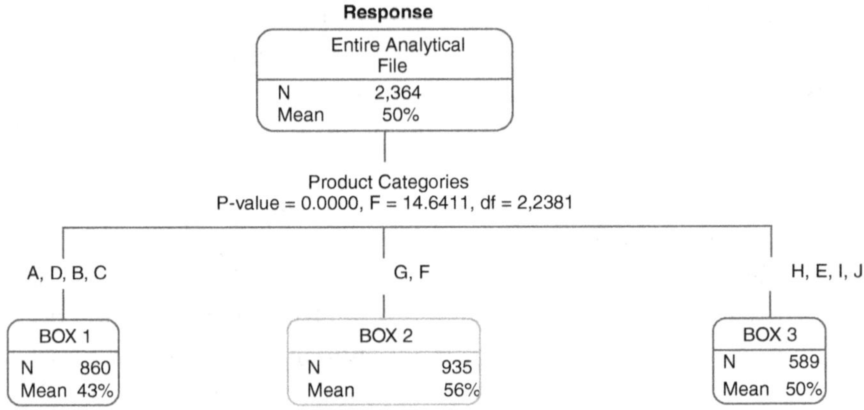

Figure 12.7 CHAID Used to Group Product Codes into Categories

This ordinality (1, 2, 3 outcome values) is determined by ranking the nodes based on the response rate mean within each node. By ordinality, the lowest response rate node (box 1: A, D, B, and C) is assigned a value of 1, the next highest response rate node (box 3: H, E, I, and J) is assigned a value of 2, and the remaining node with the highest response rate (box 2: G, F) is assigned a value of 3.

In this example, new and more meaningful variables can be derived in situations where character variables have many different outcomes. This is prevalent in the retail industry where it is not unusual to observe over a hundred different product codes or SKUs in a given company.

CHAID and factor analysis demonstrate that the creation of variables can still continue even when the analytical file has already been created. Creation of variables will continue as long as there are insights into how best to manipulate information. New information and variables are easily added to the analytical file based on the learning arising from stage 3 of the data mining process. What happens next in the process?

With the above-mentioned statistical routines, the discovery process in this stage includes routines that allow data miners to better understand the information that could comprise a given solution. This is even more necessary if data miners must communicate the results to business end users. Examples of such routines are called EDAs (Exploratory Data Analysis) reports. An example of such a report is shown in figure 12.8.

In this first example, which is an attempt to build a response model, the variable, tenure, has a strong positive relationship with response. In other words, the higher the tenure, the more likely the person is to respond. Another example of an EDA report is shown in figure 12.8.

Tenure in Years	% of Customers	Response Rate	Response Index
0–1	25.0%	2.0%	57
2–4	25.0%	3.0%	86
5–7	25.0%	4.0%	114
7+	25.0%	5.0%	143
Average	100.0%	3.5%	100
Age			
Under 25	25.0%	6.0%	171
25–35	25.0%	4.5%	128
35–50	25.0%	2.5%	71
50+	25.0%	1.0%	28
Average	100.0%	3.5%	100
Household Size			
1	25.0%	4.0%	114
2	25.0%	3.0%	86
3	25.0%	4.0%	114
4+	25.0%	3.0%	86
Average	100.0%	3.5%	100

Figure 12.8 Examples of Exploratory Data Analysis (EDA) Reports for Tenure, Age, and Household Size

In this second example, older people are less likely to respond, and this indicates the negative relationship between response and age. Meanwhile, let's take a look at the third example in figure12.8. In this example, the report indicates that there is no relationship between household size and response.

With a number of diagnostic tools already used here to develop models, such as correlation analysis, factor analysis, CHAID, and EDA reports, the analysis is now ready to employ more robust statistical routines that will create the final model. Analysts now understand the information input into the statistical techniques. Analysts must understand the information considered for a given model. Just throwing raw data and information into a model-building technique is a recipe for disaster. Given this deep understanding of information that can potentially be used as inputs into the model, a number of different techniques can be employed. Many mathematical techniques are available to analysts. In marketing, traditional techniques, such as logistic regression and linear regression, work quite well in most cases. Some of the other and more advanced techniques, such as neural nets and genetic algorithms, are other options. The difficulty with using these options is that they are very complex and difficult to understand particularly with regard to the output. Nevertheless, these options can sometimes demonstrate clearly superior results (performance lift) when compared to the more traditional techniques, and so they should be used. Data mining experience has demonstrated that many of these advanced techniques do not significantly outperform the more traditional techniques. This is because these more advanced techniques require tremendous volumes of data, and the data needs to be less noisy or have smaller portions of unexplained variation. This is as important here as in other business applications, such as optimizing the degree of quality control in a manufacturing process; there is much less noise or "unexplained" variance. The unexplained variation is small, and this suggests that much of the data can be explained. Ultimately, the more advanced mathematical techniques work against the explained data as analysts work to achieve the best solution.

However, this gets to the heart of the debate between academics and practitioners. In marketing, large random unexplained variation is a fact of life. Trying to explain away this random variation can lead to results that appear during the development of the solution and that are very good, but when they are applied in the business they lead to performance that is less than optimal. This particular situation in data mining is often referred to as overstatement of results. These techniques must be used judiciously, and this means that the use of validation samples is even more imperative. Otherwise, results achieved in a marketing campaign will always be inferior to what was achieved during development of the solution.

The more traditional techniques are less complex and don't have the capability to overstate results. This can be easily demonstrated by the typically low F values and R^2 values observed when a final model is created. Much of the actual variation in a final model is unexplained. Obtaining an R^2 value of .05 might represent an optimum model building scenario for data miners given the current data environment. However, the threshold of success is not statistical diagnostics but rather gains charts/decile charts, which will be discussed later.

With these more traditional techniques, what actually happens and how are the results communicated to end users? Users will conduct logistic or linear regression on the analytical file; that is, users define the objective function or the behavior to be optimized and the independent variables as those variables deemed to be potential model predictors of the desired behavior. A series of regression analyses are run against the data, and each regression equation represents one option given a certain set of variables. The objective in running these series of regression routines is to eliminate variables such that one final model is provided whereby the best 10–15 variables are run as the last regression equation.

Typically, it is not unusual to run 5–10 of these regression analyses prior to the development of a final solution. The number of these regression analyses depends on the learning and insights from correlation analysis, CHAID, and EDAs. The first tool to prescreen the variables is correlation analysis; analysts use a statistical significance level threshold of 95% to determine which variables should be considered in developing the final model or solution.

This group of statistically significant variables from the correlation analysis and other insights from both CHAID and the EDA reports can be used to identify variables to be used as inputs into a regression routine (iteration). However, analysts still have to find a way to group all these variables. Using correlation analysis, analysts can formulate groupings (iterations) based on the relative statistical strength of each variable. In determining what variables to include in all the groups, the decision is more scientific as analysts decide to exclude variables that are below a certain statistical confidence interval based on results from the correlation analysis. In determining the number of groupings (iterations), the solution tends to be more subjective. This subjectivity occurs when analysts determine how many variables should be in a given grouping. One rule of thumb is that no more than 15 variables should be included in any regression equation. By including all variables (such as 70 variables, which is not an unlikely situation) based on EDA and correlation results, information is lost because only one routine is performed. With 70 variables in one routine, some of the lower impact variables may be excluded from the model because of the interaction or multicollinearity with so many of the strong variables. Accordingly, analysts will group variables together based on their strength. In the initial phases, variables of comparable strength (correlation coefficient values) will be analyzed against each other within one group. This effect mitigates the impact of excluding any variable of lower impact from a final model, at least during the initial run of regressions.

The output of correlation, EDA, and CHAID, shows that the 40 remaining relevant variables from our preliminary analysis should be sorted into 4 groups. Some art can also be used to eliminate variables known to be correlated; in this way variables are extracted that have a certain level of uniqueness or independence from each other. In the above example here (figure12.9), if one regression routine is performed on all 40 variables, it is highly unlikely that any variables from group D will make it into the final model. They are excluded because of the overwhelming influence of Group A variables. By performing a series of regressions (i.e., one regression for each of the groups), the number of variables in each group is reduced based on their interaction with each other as well as on their impact on the objective function or dependant variable.

In the example in figure 12.9, the first regression is on the weakest variable groupings (Group C and Group D). The surviving variables of Group

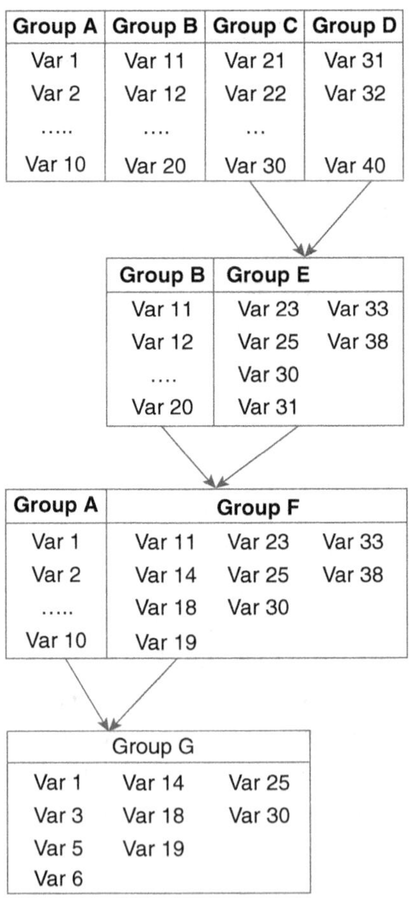

Figure 12.9 Example of Stepwise Iterations to Develop Final Model Variables

C and Group D, called Group E, are then regressed with the variables of Group B. The surviving variables Group F are then regressed with the strongest group (Group A) with the final model variables including Group G. Here, the weakest variables are analyzed first to identify survivors and to then see how they perform against the strongest variables.

With this approach, variables from the initial group D are more likely to make it into the final model. Since certain variables are excluded primarily on multicollinearity, especially with the higher groups (A and B), there are fewer variables from these stronger groups (A and B) that can compete with the lower group. In some cases, variables from the lower group (Var. 25 and Var. 30) can bring in certain information into a model that impacts the overall solution and is unique relative to the other model variables (low multicollinearity).

With this series of regressions, a final model is eventually produced. But how are the results communicated to business end users? The approach used here results in the development of a solution that is parametric. Why is it important to understand the difference between parametric and nonparametric? It is important to understand this distinction because parametric implies that each variable has a weight or coefficient assigned to it that is determined based on its impact with the modeled behavior as well as on its interaction with the other model variables (multicollinearity). In nonparametric solutions, there are no assigned weights or coefficients to each model variable. The variables themselves create the solution and are not influenced at all by multicollinearity; this suggests that there is no need for weights or coefficients to be assigned to variables. In effect, these nonparametric solutions represent a series of business rules. A good example of such a solution is CHAID, where the solution represents those optimal segments that have been determined by a series of business rules.

Meanwhile, the parametric solution is comprised of an equation with weights such as:

RESPONSE = .008 + .0015 * Income -.01 * Age.

Age and income are parameters, but the solution is parametric due to weights or coefficients assigned to each parameter (+.0015 to income and -.01 to age). The actual weight or coefficient is determined based on both the impact of the variable with the objective function, such as response, its interaction with other variables (multicollinearity), and the magnitude or scale of the variable. For example, age and income will have two widely different coefficient values. In a response model, age, varying between 0 and 100, will have a much larger coefficient than income, which might range from 0 to millions of dollars. Yet, the coefficient will also increase or decrease in absolute terms based on its impact on response where increase in absolute value implies a stronger significance for response and the reverse holding true for a decrease in absolute value. In addition, strong multicollinearity with other variables will also reduce the parameter or coefficient of a given variable.

After the equation is produced, what is presented to business users or management? The equation itself might be of interest to more technical or advanced data mining users. But to other people in the organization who have a vested interest in the model, communication of the results must be more focused on business. This business-oriented communication must focus on two areas:

- Description of the solution
- Impact on the business

The table shown in figure 12.10 represents a model with six variables or parameters and lists each variable in order of priority or significance with the model. This can be seen in the column "Contribution to Overall Equation" where the first variable (Behavior Score) accounts for 35% of the model's overall power, and the sixth variable (Are Female) accounts for 5% of the model's power. The column called "Impact on Response" describes the relationship or trend of a given variable with response. For instance, the most important variable (Behavior Score), as attested by its 35% contribution to the overall model equation's power, indicates that persons with higher behavior scores are more likely to respond. Conversely, the

Model Variable	Impact on Response	Contribution to Overall Equation
Behavior Score	Positive	35%
Average Spend	Positive	25%
Have an RRSP Product	Negative	15%
# of Financial Inst. Products	Positive	10%
Avg. % of Credit Limit Used	Positive	10%
Are Female	Negative	5%

Figure 12.10 Example of Final Model Variable Report

third strongest variable (15% contribution to overall equation) indicates that persons with an RRSP product are less likely to respond.

With this kind of information, the general characteristics of the solution can be described in business terms. For instance, the information in the table in figure 12.10 would indicate that responders have the following characteristics:

- Higher behavior scores
- Are higher spenders
- Not likely to have an RRSP
- Do have a variety of financial products
- Are heavy credit revolvers
- Are not female

Now, the same kind of approach is also used in other strategies. The key to success in using the variety of statistical techniques now commercially available is the ability to extract the relevant information that will yield meaningful business insights. In regression modeling, it is important to know the key characteristics of the model and their relationship to the objective function. Clustering solutions yield insights allowing business users to understand the key drivers or characteristics of a given cluster segment, and factor analysis provides a means to summarize or reduce data into a smaller more manageable set of variables for analysis.

Data mining is about extracting trends and meaningful business trends that can be acted on in some business initiative. Statistics allow data miners to apply science to the data to further reduce the degree of subjectivity in an analysis, but subjectivity will nevertheless always play a role in the overall solution. In fact, in most cases this subjectivity plays a significant role because it is based on the domain or business knowledge and expertise of the various key stakeholders in the given organization. In many cases, the science confirms hypotheses gained from domain knowledge.

This page intentionally left blank

CHAPTER 13

GAINS CHARTS

MANY DIFFERENT SOLUTIONS ARE PRODUCED IN DATA MINING. As previously discussed, solutions can be both of the statistical and the non-statistical variety. How is the impact of a given solution assessed? In particular, how is this impact assessed in business terms that are understandable to the business executives? The answer is through validation by what is often referred to as a gains table, an example of which is provided in figure 13.1.

Gains tables work in a very simple manner. If a solution is being produced regarding a given set of observations (this could be customer records, for example), the solution is then applied to this set of observations. The set of observations can then be ranked by the solution into groups, which are most often either deciles (10% intervals) or half deciles (5% intervals). The top group or interval would represent the group identified as being the highest priority in terms of the solution, and the lowest group or interval would represent the group with the lowest priority. Based on these intervals, the observed or actual reported metric is determined and optimized in a data mining project. For instance, in the response model in figure 13.1, the cumulative average response rate for each interval (10% deciles, for example) would be reported. The next column would report the percentage of all responders within the entire sample that is being captured in the interval. Other columns included are the cumulative response rate index, which essentially comprises the lift or the ratio of each interval's cumulative response rate relative to the average response rate for the entire sample. In addition to these columns, the return on investment (ROI) could also be included if cost per promotion effort and revenue per sale are available. The last column refers to the dollar opportunity cost, and it represents the dollars saved as a result of the data mining solution. Let's take a look at how this is calculated in the first row (0–10%). By

% of Validation Sample	Validation Names	Cumulative Response Rate	Cumulative % of Total Responders	Response Rate Lift	Interval ROI	Modelling Benefits
0–10%	20,000	3.50%	23%	233	145%	$26,667
10–20%	40,000	3.00%	40%	200	75%	$40,000
20–30%	60,000	2.75%	55%	183	58%	$50,000
30–40%	80,000	2.50%	67%	167	22%	$53,333
40–50%	100,000	2.25%	75%	150	-13%	$50,000
.	.					
.	.					
.	.					
90–100%	200,000	1.50%	100%	100	-58%	$0

Figure 13.1 Gains Chart Example

	# of responders	Promotion Costs
With Modelling	700	$20,000
Without Modelling	700	$46,667
Modelling Benefits		$26,667

Figure 13.2 Calculating the Modeling Benefits at 0–10% of the Gains Chart

determining the number of responses that were achieved with modeling in this first row (700 = .035 x 20000), we then want to determine the incremental marketing dollars that would have been spent to achieve this same response quantity but now without modeling. A schematic of how this would be calculated assuming a promotion cost of $1.00 per effort is shown in figure 13.2.

Note that the amount of $46,667 is reached by simply dividing the overall response rate of the population (1.5%) into the number of responses that are captured by the model in the top 10% (700/.015) and then multiplying by $1.00.

In evaluating a model or solution through this gains table, analysts primarily look at how well the given solution rank-orders the desired objective. In the case of a response model, response rates should optimally be rank-ordered very well from the top decile (0–10%) to the bottom decile (90%–100%). Another way to review the gains table is to plot the interval response rate of each decile with deciles on the x-axis and interval response rate on the y-axis, as is shown in figure 13.3.

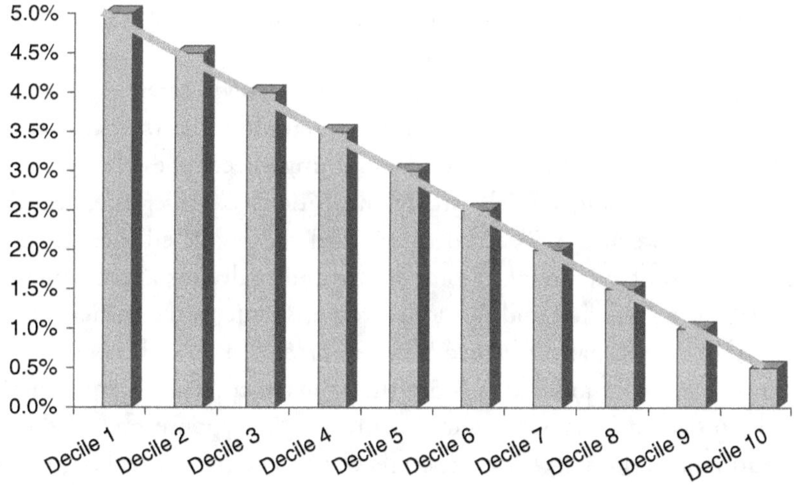

Figure 13.3 Lorenz Curve by Plotting Interval Response Rate versus Model Decile

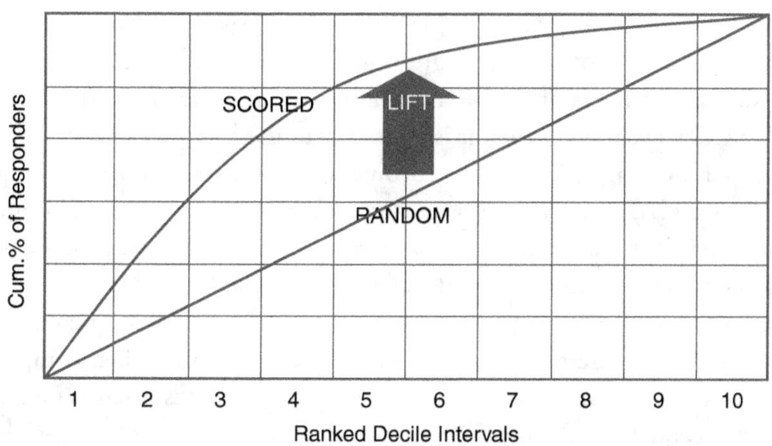

Figure 13.4 Gains Chart: Plotting the Cumulative % of Responders versus Model Decile

This is often referred to as a Lorenz curve, which represents a line that is plotted based on the midpoint of each histogram or bar in the table and the steeper the line, the better the solution. A flat line would indicate that data mining has been completely ineffective.

Another common way in which data miners evaluate solutions is to plot the cumulative percentage of the desired solution on the y-axis versus the ranked intervals on the x-axis with 1 being the top rank and 10 being the lower rank. Using the response model example, figure 13.4 shows that good solutions have a nice parabola whereas an upward straight line represents a completely ineffective solution.

To put a value on a given solution, data miners will often use this gains chart to measure the space between the straight line and the parabola and refer to this value as a measure of data mining effectiveness.

When evaluating models, many institutions consider the degree of accuracy between the observed metric and the predicted metric. In the case of response rate models, some statisticians will look at the difference between the predicted and the actual response rate as the primary focus. Although a minimal difference between predicted and observed usually results in a good model, this is not the primary criterion. Rank-ordering of the observed behavior, which in this case is response rate, is the key performance criterion. Often enough, various statistical techniques will be employed in a given solution. For example, both linear and logistic regression may be used in a particular solution. For binary outcomes such as response, theorists will argue that logistic regression should be used. Yet, if rank-ordering the behavior is the most desired solution, then using linear regression or logistic regression is equally valid. However, if the predicted outcomes of the binary solution are required as part of a more comprehensive model, then it is necessary to use logistic regression.

If rank-ordering is the desired outcome, even less robust techniques, such as CHAID, may deliver an acceptable solution. The actual output from CHAID represents a group of customers where each group can be ranked based on the observed behavior. This is less robust than techniques such as linear regression or logistic regression where the output is an individual score for each customer record. However, when comparing these techniques, in some cases even the less robust ones, such as CHAID, perform equally well in achieving the desired objective of rank-ordered behavior.

Statistical measures can be used as key indicators or metrics when comparing the performance of different statistical techniques in the development of a desired solution. These statistical diagnostics can be extremely important for statisticians and mathematicians, but the metrics can seem arcane to business executives. Business managers and executives need metrics and measures that evaluate different options in a very pragmatic manner. As seen previously with gains charts, metrics such as cost per acquisition, cost per order, cost per retained customer, and ultimately ROI represent just some of the critical key indicators for business executives and managers. All these metrics are simple spreadsheet calculations once the gains/decile charts are produced.

The challenge in communicating meaningful business results to business stakeholders is using historical information where the desired behavior can be validated. In the ideal scenario, a live historical campaign can be used as both our development and validation sample for building the model. In other words, a previous campaign or business initiative was conducted, and from that we can identify the responders and nonresponders of that campaign. Ideally, this campaign or effort was conducted with a random sample, which means the promotion was sent randomly to customers rather than based on specific list criteria. A random validation sample is critical to any modeling exercise. If the given sample is random, the results from the validation can be applied to the whole population. Conversely, if the validation sample contains only people under 30, for example, then any results and model learning from this sample can only be applied to this segment of the population.

EVALUATING SOLUTIONS

When comparing different techniques, approaches, or models, the key approach is to apply the same validation sample for each and then compare the results. If a given approach, technique, or model is superior for a given solution, then the trend line or Lorenz curve will be steeper.

In the example shown in figure 13.5, the steepest line or slope is model 3, which indicates that this solution best rank-orders the observations. As rank-ordering is our primary measure of success, model 3 clearly is the optimal solution.

As the world of data mining continues to grow, consistent methodologies for evaluating new data mining techniques become more and more important. Leading-edge practitioners and academics will continue to push the envelope in identifying solutions, software, and approaches that will

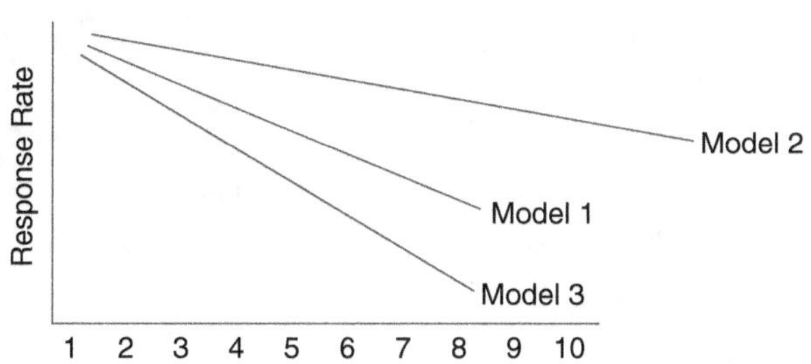

Figure 13.5 Using Lorenz curves to determine the Optimum Solution

further enhance business solutions. However, business executives involved in data mining must have a robust way of evaluating new techniques. For example, companies are often inundated with new technologies and software that purport to provide superior solutions. These datasets are commonly referred to as R&D datasets. Models and solutions have already been developed off these datasets and been applied to validation samples in order to establish a performance expectation level. New technology or software is then applied to these datasets in order to determine if and how much incremental value is provided by this new technology.

The same approach is also taken in the evaluation of new data sources. In this case, the analytical file is recreated with the new appended data. Once again, solutions with and without the new data can be compared in terms of incremental performance. This incremental improvement in performance as seen by the increased steepness or slope of the Lorenz curve can then be weighed against the cost of the new data in terms of whether or not acquisition of this data is economically justifiable.

The use of different approaches, techniques, and new data sources to arrive at an optimum data mining solution should be based on a standard evaluation yardstick. The use of a gains chart or decile chart is certainly a very good example of a yardstick for such a standard evaluation. With this kind of standard evaluation, data miners can derive the necessary insights that will allow for the successful evolution of the data mining industry.

CHAPTER 14

USING RFM AS ONE TARGETING OPTION

NOT ALL DATA MINING SOLUTIONS REQUIRE TOOLS WITH STATISTICAL analysis. Although we have focused much of our discussion on building predictive models, doing something simpler can in many cases be equally appropriate. A good example of this is the RFM index (recency, frequency, and monetary value); this index represents one nonstatistical method of targeting customers for a given business initiative. Here, the analyst can use three key pieces of information:

- Recency since last purchase
- Number of purchases in the past year (frequency)
- Average value of purchase (monetary value)

Let's take a look at an example in figure 14.1.

In this example, separate indices are calculated for recency, frequency, and average value of purchase. These three indices are calculated by dividing the average value of the entire database into the specific value for a given customer. With an index of 1 being average, values above 1 represent the above average performers, and the opposite is true for values below 1. You will note that the calculation is reversed for recency because here low values are good. The index calculations are as follows:

- Recency: 6/3 = 2
- Frequency: 4/3 = 1.33
- Average value of purchase (monetary): 150/100 = 1.5

	Recency-# of months since last purchase	Number of purchases within last year	Average value of purchase
Customer A	3	4	150
Customer Database	6	3	100
Index	2	1.33	1.5

Figure 14.1 Example of Indexes for Recency, Frequency, and Monetary Value

	Recency	Frequency	Monetary	RFM
Customer A	5	5	5	15
Customer B	4	5	3	12
Customer C	5	4	3	12
Customer D	1	2	2	5
Customer E	2	1	2	5
Customer F	1	1	1	3

Figure 14.2 Example of RFM Index Using 5 × % Matrix

The actual RFM index for Customer A would then be (2 + 1.33 + 1.5)/3 = 1.61,where we assume that each behavior (recency, frequency, and monetary behavior) are weighted equally.

Another approach is to sort customers based on each behavior (recency, frequency, and monetary value) into 5 quintiles, as shown in the example in figure 14.2.

In this example, 5 represents the best performing behavior while 1 represents the worst performing behavior. In this approach, we just simply sum up the values, and the higher the number, the better the overall behavior. In the example in figure 14.2, Customer A is the best customer with an RFM score of 15, and Customer F is the worst with an RFM score of 3. Once again, we assume each behavioral metric to be weighted equally. With this approach, Customer B and Customer C are both shown as good performers but with identical scores of 12, and Customer D and Customer E are performing less well but again with identical scores of 5. There will be many customers with ties in this approach as different combinations of behavior values can lead to the same score.

With RFM, decile reports can be created that rank order or sort customers into groups with decile 1 the best RFM group and decile 10 the worst RFM group.

% of Customers	Avg. RFM Index of interval	# of Customers
0–10%	3	10000
10%-20%	2.5	10000
20%-30%	2	10000
30%-40%	1.75	10000
....		
90%-100%	0.3	10000

Figure 14.3 Example of RFM being used as Targetting Tool to Select Customers

In the example (figure14.3), customers can then be selected for a variety of different business initiatives.

The use of RFM is not complex and very pragmatic and allows the selection of records based on prioritizing particular historical behavior. Yet, RFM is limited by the fact that it is considered reactive: decisions are based on what has happened in the past. This is not necessarily bad because the past can be an indicator of what is yet to come. But in today's world, the need to be proactive or preemptive is great as businesses want to make decisions based on what is likely to happen in the future. Analytics, using RFM variables along with other variables that are predictive by relating past behaviors to some future behavior allow organizations to be more proactive in their decision making. RFM will always remain a "quick and dirty" targeting tool, but the use of predictive analytics and predictive models will be the preferred option if solution optimization is the goal.

This page intentionally left blank

CHAPTER 15

THE USE OF MULTIVARIATE ANALYSIS TECHNIQUES

THE MOST COMMONLY USED TECHNIQUES IN PREDICTIVE MODELING ARE linear and logistic regression. The statistics in linear regression predict outcomes with a continuous range of values; logistic regression predicts outcomes that are categorical in nature. The most commonly used logistic routines are used to predict yes/no behaviors, such as response, attrition, or credit default. The outcome derived can also be a set of rules such as CHAID rather than a score, which is the predicted outcome of either logistic or linear regression.

Both logistic and linear regressions have existed for decades. However, only in the past 10 to 15 years have these techniques become common business practices in marketing as organizations strive to improve their overall targeting efforts. These techniques, as stated earlier, have been widely used for years by companies in their efforts to target customers who are creditworthy.

As with all statistical techniques, the underlying objective is to maximize explained variation and minimize unexplained variation. For example, in this equation to predict response:

$$\begin{aligned}&\text{Response} = .48\\&+ .3 \times \text{female}\\&+ .15 \times \text{number in household}\\&- .00004 \times \text{income}\\&+ .07 \times \text{number of years education}\end{aligned}$$

For now, we can set aside the academic discussion that this equation is linear and attempts to predict a probability function. Academic purists will argue that this equation should be transformed to a logistic function if we want to predict response. But as we have discussed in previous chapters, the main priority in developing business solutions is to optimize the solution's ability to rank-order or to differentiate records based on the targeted behavior. In most cases, an accurate estimate of the solution is a distant second priority. However, for now we can assume that this linear function is a good approximation of response.

In the equation above, there are four variables that indicate the following:

- Females are more likely to respond
- Persons living in larger households are more likely to respond
- Persons with higher income are less likely to respond
- Persons with higher levels of education are more likely to respond

In building an equation that maximizes variation, the variation of values among each of these variables best approximates how the response rate values vary. This logic is very similar to what was discussed regarding correlation analysis. A visual plot of this data that captures the relationship between age and spending is provided in figure 15.1.

Naturally, the line in the graph represents the predicted outcome from our equation. The points (x) represent the observed outcome. In building solutions, the goal is to minimize the distance between the observed points and the line (predicted outcome that is depicted for one point by the letter). At the same time, the results reveal that there is some trend in the connection between age and spending. The trend or line represents a plot of the

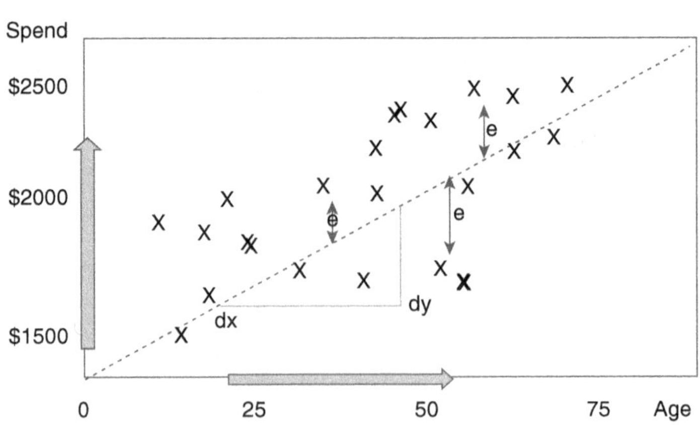

Figure 15.1 Depiction of Best Fit Line in Regression

explained variation. In this example, if spending is our targeted behavior, we observe that age and spending are trending upward; this indicates that age (i.e., older people spend more) is a good predictor of spending if spending is our targeted behavior. This so-called trend in essence represents the explained variation, which in statistical terms is often referred to as the R^2. Perfectly explained variation or a perfect equation will yield an R^2 of 1; this implies that the variation in observed response is 100% explained by the equation. In this scenario, every observed point x would be on the straight line. However, in reality and in particular in the business world, this is never the case. In fact, the level of unexplained variation is always very high and represents a real limitation to the creation of robust solutions. It is highly common to see R^2 of most equations never exceeding.05. For most data miners, this represents a real challenge since this high level of residual variation results in practices that call on both art and science.

Some of the newer techniques, such as neural nets and perhaps genetic algorithms, attempt to explain this true random variation, but this variation should be left unexplained. This situation is common among some of the newer techniques and is often referred to as "overfitting."

With overfitting, these gains charts results are even further enhanced. Once these performance expectations are laid out for the marketing team, objectives and forecasts become quite aggressive and unrealistic, and in most cases they yield inferior results compared to what was expected. The use of less robust tools, such as CHAID or regression analysis, in these cases would have delivered solutions that are more easily managed when applied in various business initiatives.

Obviously, this does not mean that these more advanced techniques should never be employed; rather, these techniques along with other approaches should always be considered options. In fact, from a statistical perspective, these techniques are very powerful, and they have been employed in applications such as fraud detection, particularly in image recognition and handwriting recognition. The key to using these techniques in business and marketing applications is to have sound validation or holdout samples. The crucial difference in marketing and in trying to predict consumer behavior is that a large portion of the variation in the data is truly random or unexplained. This means that any attempt to explain this "random" variation through the use of more advanced techniques will yield disappointing results. The necessity to use validation samples can never be overemphasized in the data mining world, especially when large random variation is the norm rather than the exception.

As seen in the section dealing with the creation of the analytical file, the very last stage of the process is the separation of this file into a development and validation sample. Solutions should always be constructed on the development sample. The solutions' performance should always be measured against the validation or holdout sample; that is, we simply apply the solution to this group of records.

Although it has been argued that is the proper procedure when building solutions, in practice this is not always necessary, especially if sample size is a mitigating constraint. The lack of data in building a model would suggest that the development and validation samples should be combined in order to have enough information to develop a proper model. Although there is some risk in developing models without a validation sample, this risk is mitigated with the less robust techniques, such as CHAID and regression analysis. From an experienced standpoint, minimal difference is observed in results when the solution is applied to a validation or development sample, particularly if the proper discipline has been exercised in model development. If the model and its variables are statistically sound, then it is expected that the results will be so as well. These simpler techniques are only producing results from that portion of the total variation that is explainable, and in most cases that portion represents a very small part of the total error. Hence, simpler techniques, such as CHAID and regression, are sufficient for developing solutions to explain this small portion of the total variation. However, this explainable error must be duplicated when validating results that are similar to what occurred in development. With the more robust techniques, some of the model's predictive capability is based on being able to explain variation that is truly random or unexplained. This attempt to "explain the unexplained" is not replicable in a validation sample, and drastically different results between development and validation are expected if a more robust and complex model were not properly trained. In my experience, when models have been properly trained with these more robust techniques, the results are indeed similar to those produced by CHAID and regression.

In the world of marketing, situations will rarely arise that allow a large portion of the variation (i.e., customer behavior) to be explained. In other words, techniques like neural nets and machine learning seldom beat the more traditional regression techniques in the world of marketing. But the small portion that can be explained results in significant increases in performance. Because these solutions are applied to hundreds of thousands of customers, these simple models can achieve tremendous lift in a given customer behavior and ultimately add hundreds of thousands of dollars in revenue to a given campaign.

Understanding that this small explainable error is the norm in database marketing can help explain why linear or logistic regression can be used to predict binary outcomes. From a statistician's perspective, this is heresy. From a data miner's perspective, it is simply a matter of what works and what doesn't.

Binary outcomes, such as defection, credit loss, or response rate, are essentially yes/no outcomes. Did a person respond? Did a person go into default? Did a person cancel?

In the world of database marketing, rudimentary use of statistics will often suffice and be preferable to the more complex approach. As discussed throughout this book, the real key to building stronger solutions is a data environment that offers much data that is reliable and provides the flexibility to manipulate this data into more meaningful fields.

Because of the binary nature of the outcomes, these models should be producing a probability function. In other words, the outcome should produce results that are between 0 and 1. Academic purists will argue that using linear regression to estimate binary outcomes is incorrect since the predicted outcomes from a linear regression routine will have values between negative infinity to positive infinity. From the practitioner's perspective, however, what matters is that the solution provides significant performance lift when compared to other alternatives. As previously mentioned, performance lift is the practitioner's primary focus. It would be nice to have a solution that is also accurate in predicting actual response rates (i.e., where the difference between the predicted response rate and the actual response rate is minimal), but this is not mandatory. Marketers and analysts desire solutions that can best differentiate or rank-order groups in their differences from the average. The example in figure 15.2 illustrates this focus.

In this example, Scenario 2 produces a model that more accurately predicts response. However, which scenario would database marketers select as a solution? Despite its obvious inability to accurately predict response, Scenario 2 still yields a much higher lift and would be the preferred choice of marketers. The ability to produce a solution that offers significant lift in performance is the preferred objective over one that provides more accurate estimates of response rate.

If lift is the primary consideration, then there is no need to have a mathematical routine, such as logistic regression, that delivers probability estimates. Once again, given the small amount of variation that can be

	Scenario 1	Scenario 1	Scenario 2	Scenario2
	Predicted Response Rate	**Actual Response Rate**	**Predicted Response Rate**	**Actual Response Rate**
Top 10% of Model	6%	10%	4%	4.50%
Random Control Cell	3%	5%	3%	3.50%

Figure 15.2 Comparing Predicted Estimates vs. Observed Estimates

explained, linear regression in most cases will yield solutions that are often very similar to logistic regression when considered from the perspective of lift.

Why should logistic regression be considered at all in a business setting? The practical response is that often models are built as part of an overall solution, such as campaign optimization or profitability. In these cases, the predicted outcomes represent probability estimates that may be used as one metric in building the overall solution. Here's an example.

$$\text{Profitability} = \text{P (Response)} \times \text{P (Approval)} \times \text{Spend} \times (1\text{-P (Cancel)})$$

In this example, profitability is determined by a person's likelihood to both respond and be approved for this offer, and the equation then determines the applicant's level of spending for a given period of time as well as likelihood of being a customer at the time of the offer.

Converting models into probability functions is not very difficult. Once users have identified the key variables in the final linear regression, they simply input the same variables into a logistic function readily available in many software packages. In fact, in building logistic or probability models, this may be the preferred approach, especially if CPU (central processing unit) and time is an issue. It is far quicker to determine the relevant variables and statistical variables from linear regression. The selection and identification of these variables yields identical results either with linear or logistic regression. Given this finding and the higher CPU requirements of logistic regression, it is preferable to use linear regression for variable discovery of the model exercise. Then, once the model variables are finalized, they are transformed into a logistic probability function as the last step in model development. In situations where solutions are built that require the integration of several models, the combination of linear regression with logistic regression provides analysts with the appropriate tools to yield the best solution with quick turnaround time. This then allows analysts to turn their focus to the specifics of the solution an organization requires.

THE IMPACT OF MULTICOLLINEARITY

Statistics play an extremely valuable role in optimizing the decision-making process. Understanding what statistics can do in relation to a business objective is critical to this process. This is certainly the case when marketers are asked to create a target group or list of customers and at the same time identify the key characteristics defining that group.

For instance, when a list of customers is generated, the objective is to obtain the best group or list of customers. Regression analysis routines perform very well in achieving this objective. The use of these routines helps to identify the characteristics that will best predict a given outcome. One real advantage of these techniques is that they also take into account multicollinearity. What is multicollinearity? A good practical example is the notion that education usually leads to higher income. In an analytical exercise such as response or attrition, the high multicollinearity of both age and income would end up causing both variables to exhibit the same trend on response or attrition. In another example, we might find that gender and tenure (length of time as a customer) have no relationship with each other. In conducting any analysis on response or attrition, we cannot assume a priori that the two variables (gender and tenure) would have the same relationship on response or attrition.

In mathematical terms, multicollinearity represents the level or amount of interaction between the independent variables in a given model equation. This means that the weights or coefficients of the final variables reflect this interaction between multiple variables and at the same time are statistically significant in the overall equation. Through regression analysis techniques, many variables will be eliminated because of multicollinearity with the final model variables, but they may still be statistically significant in a univariate fashion as seen in correlation analysis routines. This type of process is critical for optimizing the predicted outcome. The key outcome in this case is a score that is the key measure used to generate lists or groups of customers.

	Years of Education	Income
Correlation Coefficient	0.11	0.12
Confidence Interval	99.0%	99.5%

Figure 15.3 Example of Correlation Results of Education and Income vs. Response

When attempting to identify the key characteristics that define a given group, using the characteristics or variables from a regression routine may be less than optimal. For example, figure 15.3 presents a case where the goal is to optimize response rates. Both variables are statistically significant for response rate in a univariate fashion (i.e., each variable without the impact of multicollinearity as compared to response rate).

The regression equation here, however, only includes income.

Response = 0.50 + 0.00001*income

Here, the equation has retained only one of the two statistically significant variables (income). The high multicollinearity between income and education has caused education to be eliminated. However, in identifying key customer triggers or behaviors for response, marketers would want to know that both education and income are key customer characteristics in the target group.

CHAPTER 16

TRACKING AND MEASURING

DATA MINING IS MUCH MORE THAN JUST TECHNOLOGY. IT IS A step-by-step process resulting in the development and application of a solution. Yet, at the same time, this process should always yield new insights and learning that can be utilized for future campaigns. Besides the development and application of solutions, the data mining process should encompass an ongoing learning environment that provides the means for continuous business improvement. It is this stage of tracking and measurement of results that provides learning and insights for future business initiatives. In fact, leading-edge organizations devote much of their resources and time to this area. Why? Being able to measure and evaluate results as part of an ongoing process provides the insights required for continuous business improvement. However, it is equally important that these results are effectively communicated so that the information and insights are disseminated to a wide audience. The empowerment of more people through increased access to these results simply means that more minds can be focused on building solutions rather than having to rely on only a select few.

Consider how measurement and evaluation fit into the overall data mining process. The first stage (identification of the business problem/issue) deals with the collection and gathering of information that will enable data miners to identify key issues and challenges. By measuring and evaluating results from recent business initiatives and campaigns, data miners can gather the most up-to-date information to formulate these key issues and challenges that continue to arise throughout this ongoing learning process.

MEASURING ROI: GETTING STARTED WITH THE DATA

Understanding what is to be measured with a business initiative can be the most significant hurdle to effective ROI measurement. Situations where measurement reports are produced but do not produce the required results often occur in marketing. Resolving or mitigating these situations requires an approach that addresses the needs of all the stakeholders involved in this process.

Once these measurement objectives and stakeholders' needs have been identified, the next issue is to decide how measurement and tracking will be done within the current data environment. This question requires us to look at the data and to determine if the measurement objectives identified can be met within the existing data environment.

For example, how is the appropriate testing matrix set up for a given business initiative or marketing campaign? What kind of learning do marketers want to obtain and how do they evaluate the performance of a given initiative or campaign? For instance, if a targeting tool is used in a campaign, results must be obtained regarding its performance. However, it is the amount and complexity of the required learning that will dictate the design of the measurement and tracking system. For example, the simplest kind of tracking is merely determining the performance of a particular initiative. The initiative could be an event, a campaign, a type of offer, a type of communication, a targeting tool, etc. In this case, the tracking requirements are quite simple in the sense that two cell codes/groups are created. The first cell code or group is the control cell where there is no occurrence of the initiative, and the second group or test cell is the one impacted by the initiative. By comparing the performance of the two cell codes, we can then surmise the impact of the particular initiative being tested. This is rather easy if the marketer is primarily interested in only one piece of learning from a campaign. Yet, in practice, marketers are always trying to obtain as much learning as possible. In setting up a measurement matrix, complexity is not the issue here; the real issue is cost. For each given initiative, test cells must be set up. Suppose that a marketer wants to evaluate four different offers along with a control offer. In this case, five cells would have to be created.

Now suppose that marketers wanted to test four different offers along with four different communication pieces. More important, suppose they wanted to test the relationship between offer and communication. Consequently, they would need to create 16 separate cells (4 offers x 4

	Communication Piece			
Offer Type	**A (Control)**	**B**	**C**	**D**
1 (Control)	5,000	5,000	5,000	5,000
2	5,000	5,000	5,000	5,000
3	5,000	5,000	5,000	5,000
4	5,000	5,000	5,000	**200,000**

Figure 16.1 Example of Marketing Matrix for Measurement

communication types); offer type 1 would be the control offer and communication piece A the control communication piece. Figure 16.1, shows a schematic of this tracking matrix with live promotion quantities.

In any tracking scenario, there is always one cell the marketing team has to select as its winning cell. Marketers will place many of the names here since this is the cell that is expected to generate the largest ROI. In the example here, the winning cell is D4 while the other 15 cells essentially represent the investment costs of promoting to 75,000 names or to 5,000 names per cell. This can be considered a significant investment since 27% (75,000/275,000) of the promotion quantity is devoted to learning. At a price of $1.00 per piece, this can translate to an investment of $75,000 devoted to learning.

Suppose marketers wanted to introduce other factors into the mix, such as region (4 groups) and targeting (4 groups); the tracking matrix in the previous example would then grow to 256 (4 offers x 4 communications x 4 regions x 4 target groups) cell matrix. Obviously, this becomes unwieldy.

This is where the use of statistics such as conjoint analysis or analysis of variance (ANOVA) can be used to reduce the number of test cells to a more manageable number. These techniques are useful in providing a scientific approach to the interaction between factors. If we assume that conjoint analysis/ANOVA demonstrates that there is no interaction between targeting and region and the other factors, then, the matrix shown in figure 16.2, could be produced.

Through statistical analysis, the number of cells has been reduced from 256 to 24. This is another illustration of the role of statistics in database marketing programs. Yet, in this scenario, the objective of statistics is not the creation of targeting tools but rather the determination of the cell/groups that need to be tested in a given marketing campaign.

	Communication Piece			
Offer Type	**A (Control)**	**B**	**C**	**D**
1 (Control)	X	X	X	X
2	X	X	X	X
3	X	X	X	X
4	X	X	X	X
Region				
Maritimes	X			
PQ	X			
Ontario	X			
Rest of Canada	X			
Targeting				
Segment 1	X			
Segment 2	X			
Segment 3	X			
Segment 4	X			

Figure 16.2 Example of Marketing Matrix for Measurement (More Complex)
Obs. Each region cell and targeting cell get control communication and control offer

Digital marketing allows marketing executives to explore all kinds of different test scenarios because the cost to create different page views and different types of e-mail messages is negligible. Although it is now easy and cheap to create these complex measurement scenarios, the limiting factor is still the human capacity to derive meaningful learning from the multitude of tests that could be designed for a particular marketing initiative. Using techniques of the ANOVA type and factorial design to reduce factors for testing accelerates marketing analysis to a new level where analyses reveal more meaningful insights in a more timely manner.

Another issue that often arises when designing a testing matrix is the question of what sample size to allot to each cell. There are a variety of factors that impact sample size. Examples of these factors that influence sample size are the overall performance rate, error range, and confidence level. The actual formula can be written as follows:

$$\frac{(\text{Confidence level}) \times (\text{Confidence Level}) \times (\text{Performance Rate}) \times (1-\text{Performance Rate})}{\text{Error Range} \times \text{Error Rate}}$$

Here are some examples to obtain clearer insight into how this formula works. For example, if we assume a response rate of 1% (performance rate) and a confidence level of 95%, which translates to a statistical z-score of 1.96, and that the results are statistically significant, and if given initiatives are 0.2% (error range) different from the mean, then the minimum required sample size is simply:

$$\frac{1.96 \times 1.96 \times (.01) \times (.99)}{.002 \times .002} = 9508$$

If the allowable range of error is increased to 0.4%, then intuitively one would expect the minimum required sample size to decrease. In fact, it does decrease to 2,377. Increases in the confidence level will always increase the sample size. For instance, if we increase the confidence level from 95% to 99% or the z-score from 1.96 to 2.58, the sample size increases from 9,508 to 16,475. Meanwhile, a performance rate of 50% always yields the largest sample size since any other combination of performance rate and its complement (1-performance rate) will always yield a smaller number in the numerator.

This formula can be plugged into any spreadsheet application. It has tremendous utility for marketers because it provides a range of required sample sizes based on the differing assumptions of performance rate, range of error, and confidence level.

Let's suppose marketers want to be more granular in their evaluation of results by determining the optimization of a given business objective. This scenario is especially applicable to targeting tools that are used in a marketing campaign. With a control or random cell set aside and then used in comparison to the targeted group, marketers can quickly ascertain if the model is effective by checking whether the results are superior to those of the control group. But is this procedure optimal? Is the targeting performing as observed when the tool was built? This is where a comparison between that status when the model was built and its current application must be created. By checking the performance results of the model when it was initially developed, marketers then have a benchmark for comparing subsequent results of targeting applications.

For instance, suppose a response model is built whereby customers are ranked into 10 deciles with decile 1 being the highest scored and decile 10 being the worst scored. The observed response rate from the validation sample for model development is then reported for each decile and plotted as a line or curve. This line or curve is often referred to as a Lorenz curve. The objective of any model or targeting is to maximize the slope of the Lorenz curve. A flat line represents a complete failure while a vertical line represents perfection.

Once this model is applied to a current campaign, the exercise can be conducted and a Lorenz curve can be created for the current campaign. This Lorenz curve is then compared to the Lorenz curve that was observed during model development. If the slopes of the two curves are identical, then the model is performing optimally. If the slope of the campaign is less than the slope for model development, then the model is not performing optimally. The key business decision in this type of situation, though, is to determine whether or not a company can live with this suboptimal performance or must invest in the development of a new model.

The following examples will serve to illustrate this concept. Figure 16.3 shows an example of two Lorenz curves; the first one reports the results during model development (validation), and the second curve reports the results after the model was applied in a campaign (campaign).

Even though the overall campaign's response rate is performing much worse than what was observed during model development, the targeting tool is performing effectively because its ability to rank-order response rate has remained more or less unchanged between development of the model (validation) and its current application (campaign). Since the slope of the two lines is identical, it can be concluded that the model's current performance is optimal. In this particular case, it was critical to evaluate the

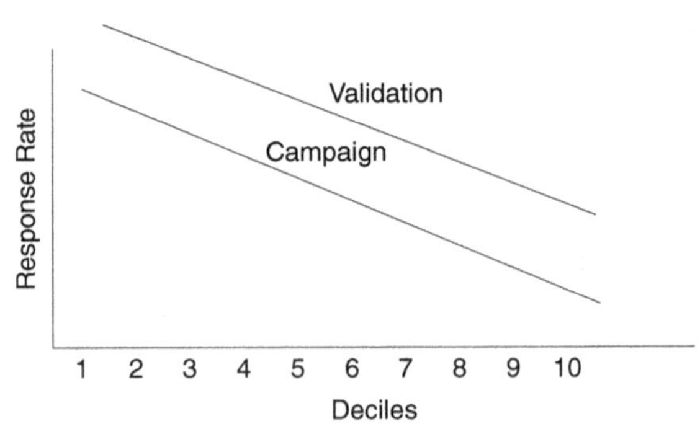

Figure 16.3 Example of Model Evaluation in a given Campaign

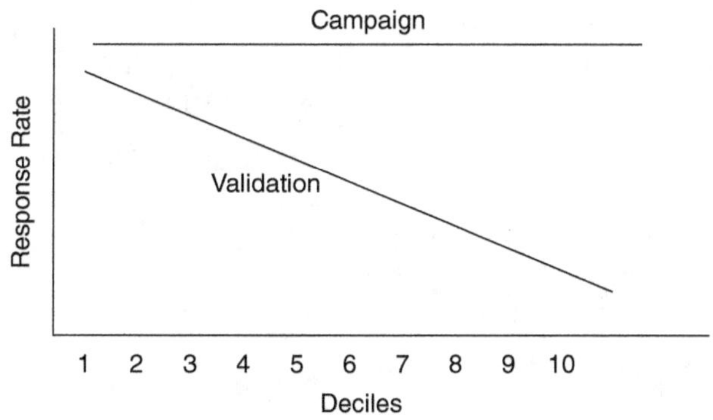

Figure 16.4 Example of Model Not Performing

model's performance in a granular manner since the first inclination of the business team might be to blame the model for the campaign's poor performance.

Figure 16.4 shows another example of reviewing a model's performance during its current application. Observe that the overall campaign is performing much better than what was observed during model development. In this particular case, the first inclination might be to attribute some of the campaign's success to the model when in fact the model did not contribute to the campaign's success. Despite the apparent overall success of the campaign, the model was a failure due to the fact that there is no rank-ordering of response rate by decile as demonstrated by the campaign's flat Lorenz curve. The slope of the campaign's Lorenz curve in this case is almost nonexistent. The business decision in this case would be rather simple in that a new model definitely must be developed.

These are just some examples of what one might consider in designing a measurement and tracking system. The key to creating any measurement and tracking system from a tactical standpoint is to establish priorities and objectives and plan accordingly. Yet, this planning can be irrelevant if practitioners do not properly understand how to use the data in the existing information environment and, more important, how to use it in a cost-effective manner.

Another way to measure performance of a model is through what is commonly referred to as the lift curve; an example is shown in figure 16.5.

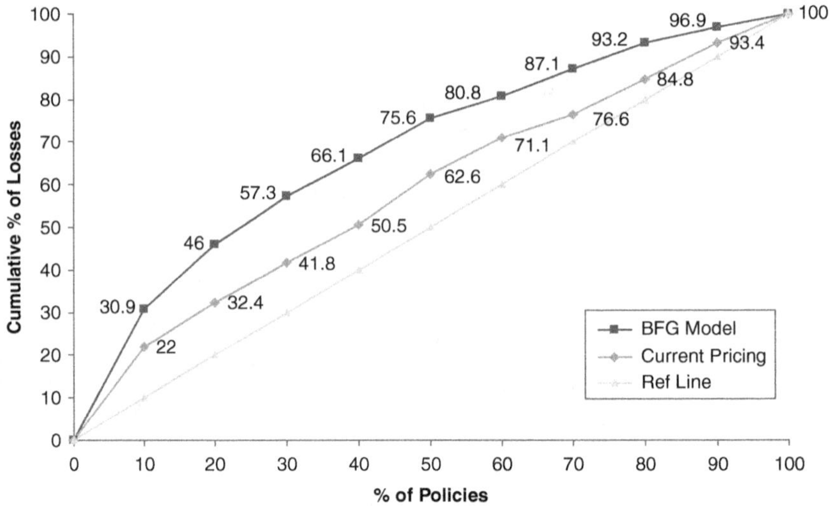

Figure 16.5 Premium versus Homeowners Loss Model Comparison of Cumulative % of Losses

In the example in figure 16.5 the objective of the predictive analytics model is to effectively target homeowner losses, and the analyst is tracking the impact of the (BFG model) as compared to the status quo (current pricing). The current way of pricing does an adequate job of identifying losses when compared to doing nothing, which is represented by the straight line. Yet, the predictive model (BFG) does significantly better, as depicted by the area under the curve between the BFG model and the current pricing technique.

CHAPTER 17

IMPLEMENTATION AND TRACKING

ONCE THE SOLUTION IS COMPLETE, THE NEXT STEP IS TO IMPLEMENT it as part of a business initiative. In some cases, these initiatives could be unrelated to marketing. For example, the development of credit-risk models could be applied in the operations area; the output of these models consists of the risk segments that are associated with these customers.

In applying solutions, the most important consideration is to ensure that the solution is being applied correctly. Initially, this requires that data quality checks will be conducted on several records to ensure that the solution has integrity on the records. This is particularly important if a third party is implementing the solution. Both the solution developer (data miner) and third party would implement the model separately using their own tools. The solution output (model variable means, model scores, etc.) should line up 100% between the data miner and third party.

Another consideration is to ensure that the information environment has not changed substantially between the time when the solution was developed and the time when it is implemented. This can be done by creating frequency distributions of key elements in the solution. A model could be examined by comparing the score distribution ranges between time of development and time of implementation, as shown in the example in figure 17.1.

In this example, the score ranges have changed quite drastically between the time of development and the most recent application. Forging ahead with this solution without understanding these score discrepancies invites failure. Investigation and analysis needs to be conducted on the database to understand why these score discrepancies exist. One method is to

% of List	Minimum Score (development)	Minimum Score (application)
0–10%	0.08	0.04
10–20%	0.07	0.03
20–30%	0.06	0.02
30–40%	0.05	0.01
40–50%	0.04	0.004
etc.		

Figure 17.1 Example of Validating Implementation Results

compare the frequency distribution results of the model variables between the time of model development and that of the current implementation.

Once the data miner is comfortable that the solution is being correctly applied within the current business initiative, he or she needs to create a testing and tracking environment in order to evaluate the impact of this solution on a business initiative.

As we observed in the preceding chapter, the intention in designing a marketing matrix is to maximize results while also maximizing learning that can be used to make better decisions in future business initiatives. In this case, key pieces of learning would involve testing how well the model performs and determining the effectiveness of various communication pieces or offers. As stated in the preceding chapter, one cell of the spreadsheet will represent the winning cell, representing those names where the model is being fully utilized to deliver its impact, as well as being promoted with the expected winning offer and the expected winning communication piece. Implementation is a detailed process. It is important to devote time and effort to ensuring that the solution is correct and makes sense in today's information environment. This time commitment is also necessary to set up the proper testing and tracking conditions for evaluating performance.

CHAPTER 18

VALUE-BASED SEGMENTATION AND THE USE OF CHAID

THE PROCESS OF MEASURING AND EVALUATING RESULTS CAN ALSO BE used to develop an overall segmentation strategy. The most commonly used approach in developing a segmentation strategy is based on value. The key and ultimately the largest challenge in this type of exercise is the identification of metrics and measures that comprise value. Once these measures and metrics are identified, a value result is built for each customer record. The results of this exercise are presented in the decile report shown in figure 18.1.

In this decile report, customers are ranked in descending order by value and placed into ten deciles. It is at this point that a judgment must be made to decide on the value segment breaks; this judgment is based on the customers' percentage contribution to overall value. In this report, the top 20% (high-value group) contributes 43% of the total value of the entire customer base. The medium-value group (20%–60%) contributes 43% of the total value of the customer base, and the bottom 60%–100% (low-value group) contributes only 14% of the total value of the customer base. In this example, the customer base has been stratified into three value segments. The value segments could also be stratified into 5 to 10 groups. The degree of stratification will depend on the number of customer records and the richness of data. For instance, a customer base of 50,000 records needs only three value segments, but a customer database of 1 million records could use 5 to 8 customer value segments. Whether to establish 5 or 8 segments for 1 million records will depend on the richness of data concerning demographics and transactions. The richness of data will determine whether meaningful profiles can be developed for

% of Customers Ranked by Value	Value Segment	Average Value	% of Value in Interval
0–10%	High	2400	24%
10%-20%	High	1900	19%
20%-30%	Medium	1300	13%
30%-40%	Medium	1100	11%
40%-50%	Medium	1000	10%
50%-60%	Medium	900	9%
60%-70%	Low	600	6%
70%-80%	Low	400	4%
80%-90%	Low	300	3%
90%-100%	Low	100	1%

Figure 18.1 Value Segmentation Decile Report

each value segment. In other words, do these profiles have some distinct or unique characteristics when compared to each other? The number of value segments should be based on the number of meaningful profiles that can be produced.

The other dimension to utilize when creating segments is one based on behavior. Besides looking at the customer's worth or value to the company, it is also important to look at a customer's recent behavioral trends that could reflect a change in that customer's behavior. This type of perspective allows marketers to design a program that is based on the customer's worth to the company and that also deals with the trigger points of this customer's changed behavior. Determining the behavior segments is an analytical exercise unto itself. The first task in this exercise is to determine an appropriate period in which to define behavior. This period typically reflects the purchase cycle of the average customer. In other words, does the average customer typically purchase in one month, two months, three months, etc.? Once this is determined, the analyst then needs to define the behavior change within this purchase cycle for growers, decliners, and stables. Behavior change needs to be calculated in terms of how purchases have changed between one period (pre) and a subsequent period (post). If the purchase period represents a time interval of six months, then one example of a pre/post period would be January to June 2013 as the pre-period and July to December 2013 as the post-period.

Once these pre- and post-periods are determined, then analysis is done to look at how percentage purchase change distributes across the entire customer database. Customers exhibiting high positive increases in purchase change are assigned as growers while high negative decreases in purchase change are assigned as decliners with the rest assigned as stable as long as they had some purchase activity in that period. The actual cutoff point in determining these behavior thresholds (growers, stables, and decliners) is somewhat arbitrary as it represents a collaborative effort between the marketer and data miner whereby both parties would have looked at the distribution in order to arrive at these cutoffs. The other segments (defectors, inactives, and reactivators) are more straightforward as they are defined upfront without any analysis.

- Defectors: Customers with activity in a pre-period and without activity in a post-period
- Reactivators: Customers without activity in a pre-period and with activity in a post-period
- Inactive: Customers without activity in a pre-period and without activity in a post-period

Once all these behavior segments are defined, a matrix can be produced that integrates both customer value and behavior as seen in figure 18.2. This matrix allows marketing decision makers to select names based on both value and behavior (value-behavior segmentation).

A second form of segmentation results from the exercises or projects that relate to building predictive models. Often enough, it may make sense not to model the entire universe for a given marketing program. As noted in previous chapters, certain variables are overwhelmingly powerful. Results would indicate that segments should be identified up front with models being built for those segments. Correlation analysis and a quick stepwise regression on key variables would reveal that one variable is clearly overwhelming when compared to other variables.

In our example in figure 18.3, correlation and stepwise results reveal that tenure is clearly overwhelming in its impact on the objective function—in this case, response.

The results in the example in figure 18.3 indicate that the tenure correlation coefficient as well as its percentage contribution to an overall equation is drastically superior to other variables that are being considered as potential model variables. CHAID (Chi-square Automatic Interaction Detector) would then be used to determine the actual ranges of tenure. These tenure ranges would then be used as segments, and models

Value	Decline	Defector	Grower	Inactive	Reactivation	Stable
High						
Medium						
Low						

Figure 18.2 Value-Behaviour Based Segmentation Matrix

Variable	Correlation Coefficient	Confidence Interval	% contribution to Model
Tenure	0.5	99%	90%
Total Spend	0.2	99%	5%
Income	-0.19	99%	3%
Credit Score	0.18	99%	2%

Figure 18.3 Using Correlation and Stepwise to Determine Variables for Segmentation

	Segment 1	Segment 2	Segment 3
Response Rate	2.5%	8%	4%
# of Names	50,000	300,000	500,000
Tenure	<1	1–3 yrs	3 yrs+
Model	1	2	3

Figure 18.4 Example of CHAID used to Define Key Customer Segments

would then be developed for each segment, as shown in the example in figure 18.4.

Here the use of CHAID to define tenure as a key segmentation variable is further supported by both correlation and regression analysis and provides another perspective on segmentation. Unlike cluster analysis, where segments are defined homogenously and in an unsupervised learning fashion, this type of segmentation adopts a supervised learning approach with response rate essentially supervising the creation of these segments.

We have discussed the use of decision-tree tools throughout this book; they warrant much discussion because they offer great benefits for data mining. First of all, let's understand what decision trees are all about.

Most of these tree-like tools employ either CHAID (Chi-Square Automatic Interaction Detector) analysis or CART (Classification and Regression Tree) analysis. Both tools again employ the fundamental concept underlying all of statistics; that is, they try to uncover characteristics that vary significantly from a given average. The main difference between the two tools is that the desired behavior or objective function behind CHAID is a binary type function (0 or 1), such as response, while for CART analysis, the objective function is continuous, such as spending or income.

Decision trees are tools where the actual output resembles a tree-like diagram. Figure 18.5 shows an example from a financial institution regarding an RRSP mailing.

In this example, the top node or branch represents the entire sample or population being analyzed. In decision-tree tools, the analysis provides two main deliverables. First, it indicates the key characteristics or variables that are statistically significant in relation to the desired behavior or response as observed in this particular example. The statistically significant variables are represented by the branches. The second main deliverable is that the analysis shows the optimum breaks or nodes within each

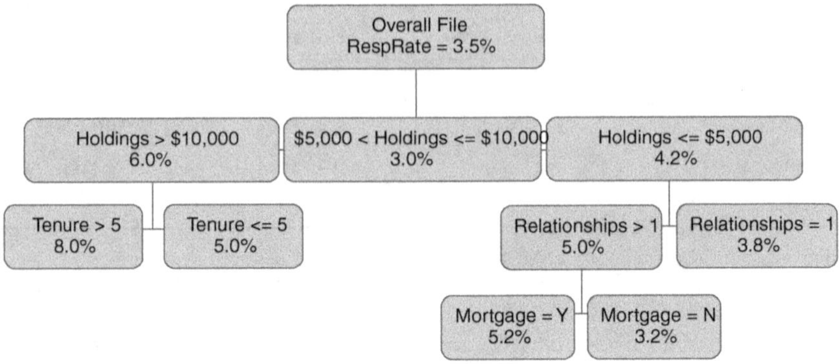

Figure 18.5 Example of CHAID as Targeting Tool

variable or branch. Note that the branches and nodes in the example here are determined through chi-square routines.

In our example in figure 18.5, the objective is to optimize response to an RRSP mailing; that is, the desired or targeted behavior is binary (response). The first branch represents holdings and is the most statistically significant characteristic against response. The nodes of this first branch of holdings are >10K holdings, between 5K and 10K holdings, and holdings <= 5K; these are all determined based on the CHAID routine. Once again, the algorithm or statistical routine determines these breaks or nodes for the user. The CHAID analysis is an iterative routine and then continues to look for the next significant variable within each node. In our example in figure 18.5, the node containing holdings >10K reveals that the next most significant node is tenure with the following breaks or nodes: less than or equal to 5 years tenure and over 5 years tenure. Meanwhile, the node containing holdings < = 5k reveals that the next significant node is relationships with breaks of 1 relationship and more than 1 relationship. Meanwhile, the holdings node with 5k to10k contains no subsequent node as there is no other variable that is statistically significant from a chi-square perspective at this particular branch of the analysis. The analysis continues its iteration where only one more branch is identified. In this case, the node representing relationships >1 has a subsequent branch related to whether or not the individual has a mortgage. The decision tree is complete at this point based on the statistical thresholds set by the user.

These kinds of routines allow the user tremendous flexibility in terms of whether the tree is going to be an exhaustive one with many nodes and branches or a simple one with few branches and nodes. Through the use of quasi "confidence intervals," such as Bonferroni adjustments as seen in the SPSS (Statistical Package for Social Sciences) CHAID tool and sample size minimum volumes, users can adjust the tree based on their own preferences. Using a restrictive minimum sample size or increasing the cutoff threshold of the Bonferroni adjustment results in fewer variables that are likely to be significant. This creates a simpler tree with fewer branches and nodes. A more exhaustive type of tree would be produced by relaxing any of these assumptions regarding sample size or the statistical threshold.

Both routines (CHAID and CART) can be used as targeting routines much like the more traditional multiple regression and logistic regression routines. As targeted routines, the actual end nodes or segments are rank-ordered from top to bottom based on how these end nodes or segments perform in relation to the desired behavior, which in our example is response. In our example, we have six segments or end nodes. A gains table or decile table can be produced using these six segments where the records are ranked by the segments' response rate performance. In our example in figure 18.5, the segments or end nodes would be ranked as follows:

- Rank 1: Holdings > $ 10K and > 5 years tenure
- Rank 2: Holdings <=5K and with relationships > 1 and who have a mortgage
- Rank 3: Holdings > $ 10K and <= 5 years tenure
- Rank 4: <=5K and with relationships =1
- Rank 5: Holdings <=5K and with relationships > 1 and who do not have a mortgage
- Rank 6: Holdings between 5K and 10K

In our example here, the use of 6 ranks to create 10 decile groups will cause many records to have the same output result. This loss of granularity in differentiating records based on response rate performance leads to decisions that are less than optimal in terms of achieving the desired business objective. The more traditional regression techniques do not encounter this phenomenon because their output are scores where much fewer records are likely to have the same score. One option for businesses is to obtain more granular results from this decision tree by relaxing the statistical thresholds and thus obtaining more nodes and ranks. The negative implication of this approach is that these increased end nodes are less statistically reliable. Users must weigh the loss of statistical reliability against increased granularity. As discussed throughout this book, business knowledge provides the additional insights to help users make a more informed and balanced decision

CHAPTER 19

BLACK BOX ANALYTICS

ORGANIZATIONS ARE NOW AWARE OF THE SIGNIFICANT CONTRIBUTION that predictive analytics brings to the bottom line. Solutions that were once viewed with skepticism and doubts are now embraced by many organizations. Yet, from a practitioner's perspective, some solutions are easier to understand than others. The challenge for the practitioner is to communicate with the business end user in such a way that the end user knows how to use and apply the solution. What does this mean? From the business end user's standpoint, this means two things:

1. The ability to use the solution in order to attain its intended business benefits
2. Understanding the business inputs

Regarding the first point, the use of decile reports and gains charts has been discussed in earlier chapters in this book; these reports and charts provide the necessary information for making decisions based on these solutions. The ability to rank order records with a given solution yields tremendous flexibility in understanding the profit implications of certain business scenarios. As businesses increasingly adopt predictive analytics, decile reports and gains charts are being recognized as the critical solution outputs in providing the necessary information for key business decisions.

Regarding the second point above, understanding the business inputs can be a challenge because here the analyst attempts to "get under the hood" of a solution and demonstrate what the key business inputs to the model are. Yet, as predictive analytics have evolved, so have the techniques. Highly advanced mathematical techniques using approaches related to artificial intelligence, which include the ability to detect nonlinear patterns, present real challenges when analysts try to explain the key components of a model. What do we mean by this? Suppose there is a distribution where a variable, such as tenure, has a trimodal distribution (3 mode points within the distribution) with response. How can one meaningfully explain the trend to business users other than telling them to simply rely on the output as presented in the overall equation? Yet, in many of these advanced techniques, there are a number of variables that exhibit this high degree of nonlinear complexity. With some of these advanced techniques and software, equations are not even the solution output. Instead, the practitioner is presented with output in the form of business rules. This scenario is further complicated by the fact that many of these nonlinear variables may have interactions with each other in such a way that the interaction between multiple nonlinear variables represents an input to the solution. These types of complex inputs pose real difficulties to practitioners trying to educate end users on what the model or solution is comprised of. The practitioner's typical explanation is that the solution is a "black box."

Yet, aside from the communication challenge of black-box solutions, the second challenge is credibility. Will the business community really trust that the equations and variables in black-box solutions are simply superior in explaining variation than in the more traditional linear and log-linear techniques? Mathematically, it may be possible to explain how a given input best explains the variation. But can this be explained in a business sense? For example, if there is a linear relationship between tenure and response, it is far easier to explain that higher tenured people are more likely to respond, which explains why tenure has a positive coefficient in a given model equation. Conversely, assuming a linear relationship between income and response, users may conclude that people with higher income are less likely to respond, and this explains the negative coefficient of income in the overall model equation.

However, explaining the impact of tenure on response becomes more difficult if the trend is a curvilinear one with multiple modes within the distribution. Tenure in this case would be captured in some kind of complex polynomial function that is very difficult to explain in business terms. This situation can even be more complicated if the solution is derived from machine learning because in that case the resulting business rules are not concerned with trying to explain an overall trend of tenure with response. Instead, the tenure relationship is captured by attempting to optimize its relationship with response along particular points of a distribution where the plotted points of tenure versus response will exhibit many nonlinear relationships.

With these types of black-box solutions, user-driven parameters provide the necessary flexibility allowing the practitioner to deliver a variety of solutions. The practitioner can alter the parameters in such a way as to deliver an almost "perfect" solution since these high-end mathematical solutions will attempt to explain all the variations, even those that are truly random. Of course, this is the fundamental problem with black-box solutions and hence the dire need for validation.

The key in assessing the superiority of one solution over another is validation and the resulting gains charts that show how well one solution predicts behavior compared to another solution. Regarding the need for a robust validation environment practitioners and academics are in complete agreement.

Yet, the inability to explain key business inputs can present a barrier to broader acceptance of these tools throughout the organization. Businesses need to have a deeper understanding of the inputs because of the measurement process used to evaluate the performance of a given solution. This process cannot occur effectively if there is a gap in understanding of what went into the solution. This is not a problem if the solution works perfectly. But what happens when solutions do not work? Under such conditions, in-depth examination is required to understand what worked and what did not work, and this certainly includes a deeper understanding of the key business inputs. This comprehension gap affecting the ability to effectively measure black-box solutions is why most organizations opt for solutions of the simpler linear or logistic type. In this type of environment, the notion of less is more is very applicable since these solutions that are not of the black box kind are more easily understood and, more important, can be analyzed in a very detailed manner even when these solutions fail to produce the expected results.

Nevertheless, black-box solutions should not be entirely excluded from the analyst's toolkit. Practitioners focus on quantitative results and therefore will always consider the results of various black-box options in the validation exercise. They will want to compare the results of these solutions to those more traditional techniques. If validation reveals that the traditional techniques are yielding basically the same results as black-box solutions, then analysts will usually opt for the more traditional techniques because they are easier to explain and use. This scenario is typically the result in most predictive analytics exercises that ultimately lead to more traditional techniques being employed. However, if black-box analytics are delivering superior results, analysts need to understand the dollar amounts that might be compromised if the more traditional techniques are used. In that case, it is best to conduct testing of both approaches in a live marketing campaign. Even better, back testing could be employed if there are relevant historical marketing campaigns or past business initiatives in which to test both approaches. The point is not to preclude different approaches but rather to exercise prudence when adopting black-box solutions that are not easy to understand.

CHAPTER 20

DIGITAL ANALYTICS: A DATA MINER'S PERSPECTIVE

THE WORLD OF MARKETING HAS UNDERGONE A FUNDAMENTAL SHIFT IN the past 15 years with the advent of digital marketing and social media as new forums for interacting with customers. Digital technology now allows marketers to communicate in many different ways. Consumers are no longer merely reacting to communications; digital technology has enhanced their ability to be proactive regarding their desires and expectations for communications they receive. Consumers now have much more control of the marketing landscape. Companies that ignore this new paradigm do so at their own risk. E-mail messages, social media communication, mobile texting, GPS alerts, and blogging just represent a few of the many different facets of communication that did not yet exist in the 1990s. Moreover, consumers can choose to interact with a company's website.

All these different facets of communication are one-to-one communication, and for many direct marketers this is familiar territory. The ability to use information and to act on that information has always been the overarching goal of direct marketers. On the basis of this goal, direct marketers have identified three pillars or principles that are the keys to success in direct marketing and also in digital marketing. These three principles are:

- List (what is the target audience of my communication?)
- Offer (what is the product or service that will fulfill a need for consumers?)
- Message (what should be the content of my communication to consumers?)

Of the three principles, the most important is the list or target audience. In fact, analytics and data mining efforts are primarily focused on this area as seen in the growth of predictive modeling and advanced forms of segmentation, such as clustering. However, we will first talk about these concepts regarding e-mail.

ANALYTICS IN THE WORLD OF E-MAIL

Capturing information concerning e-mail is similar to the process for direct mail, but it is much more automated for e-mail. Marketers can track who received the e-mail, when it was sent, and the type of e-mail, for example, in an exercise of grouping e-mails into cell code categories. On the response side, response activity can be measured based on who opened up the e-mail (open rate) and who responded to a call to action, such as a URL embedded in the e-mail. Responses to this call to action would represent the click-through rate. In addition to the increased automation of data capture for e-mail, the other advantage of digital marketing is that information is instantaneous because, in theory, all e-mail recipients receive the e-mail simultaneously once it is sent. Because information on the Internet can be accessed instantaneously, tasks can be conducted in real time. From an analytics perspective, this includes real-time reporting and real-time analysis. Once the analytical or reporting objectives have been defined, there is no longer any bottleneck or time lag; no more waiting for the data to analyze.

Given the real-time nature of information on the web and its importance to businesses for more effective decision making, pioneers of web development and marketing provide tools for tracking online customer behavior. The early adopters or pioneers developed reports that allowed companies to track such key online metrics as open rates, click-through rates, bounce-back rates, and conversion rates. From an e-mail standpoint, success is determined immediately when an organization can maximize open rates, click-through rates, and conversion rates while minimizing bounce-back rates. Conversion rates, however, may not always be accurate, particularly if customers choose to buy through a different channel.

Although the above-mentioned metrics do provide an overall glimpse of whether a campaign is successful, the ability to conduct deeper analytics is somewhat limited. Direct marketers are always interested in customers' online behavior prior to the campaign, such as the recency of the last e-mail sent, frequency of e-mails, and type of e-mails. More important, they are interested in how customers' prior online e-mail behavior will impact online behavior during the current campaign. Yet, for many organizations this kind of information must be captured in some kind of marketing database; this is again familiar territory to experienced direct marketers but not necessarily to specialists in the online world. For many organizations, marketing databases are definitely not the norm. Yet, such databases provide rich historical information that will yield valuable learning concerning the online trends and behaviors of customers and their impact on future e-mail behavior.

Analyzing Web Browsing Activity

Next, it is also important to consider the reactive side of online marketing, such as individuals browsing a company's website. Browsing a company's website is very similar to viewing a TV or billboard ad except that this browsing behavior provides readily available information to a company. The first issue is how to gather this information effectively. This implies that customers should be effectively tagged as unique when they arrive at a web site.

In the early days of the web, cookies or tags were created based on the IP (Internet protocol) address of the computer. This cookie or tag represented the unique identifier of a given customer. This system had flaws or complications that limited the ability to identify real unique customers. For example, multiple computers could have the same IP address, unique customers could also be using multiple computers with different IP addresses, and IP addresses were dynamic, which means that these addresses were changing at certain time intervals determined by the Internet service provider. As a result, a person using the same computer at different times could be considered as two unique users.

Today, most companies gathering information at the customer level utilize a user authentication system that asks for name and address and perhaps some other demographic information. Customers also have to enter a password at the first point of contact. Subsequent visits to the site will simply ask for this password, which then maps this visit to the correct unique customer. This information can then be overlaid on the customer database to determine if this visitor represents an existing customer or prospect. Marketers need to provide some incentive in order to get customers to identify themselves here. In many cases, loyalty or warranty programs are offered as one means of incentivizing customers to identify themselves. With a better ability to recognize unique visitors to a given site, the next issue is to understand this information and specifically the source information. The source information of visitors to a website is contained in log files that contain the following information:

- User ID
- Time of login
- Click/page request
- Status code, that is, the outcome of the request
- Size of the object referred to the user
- Referrer (where the user came from)
- User agent browser

The status code and user agent are probably the least important pieces of useful information in web analytics. For people with much experience working with data in the offline world, log files resemble transaction or billing files that are seen in traditional database legacy systems. RFM (Recency, Frequency, Monetary Value) variables as well as transaction type variables that are created in typical transaction files can be created from log files and can be represented as:

- Recency of last login
- Number of logins in a certain time period.
- Types of page views
- Page view behavior over periods of time.

Again, the limitation for many current web analytics providers is the inability to provide a longitudinal or historical view of the online customer's behavior. Obviously, a web analytics package bought off the shelf can be very good at evaluating behavior in terms of clicks on web pages and number of clicks over a certain time period. But such software cannot look at the online customer's behavior prior to the actual campaign. As in the e-mail example, a marketing database or data mart is the required tool that can provide this capability. With a marketing database and information about the history of the online customer, marketers can identify page view trends, the volume of log activity, and the recency of log-in activity and observe how this will impact future online behavior. Marketers are accustomed to marketing databases providing this historical view of the customer. This is clearly the analytical challenge for marketers who conduct much of their marketing activities online and on the web. Yet, direct marketing experience and expertise continues to provide the necessary thought leadership for optimizing data use.

Log file data provides rich information concerning page view behavior as well as recency and frequency of visits to a given site. Additional pieces of information provided by web data that are unique to this environment include the length of time a visitor spends on a given site. With all this rich information, models can be built to predict the propensity of any click behavior, including a page view, an order, a response to a survey, and a link to another site. Since time of engagement on the web is captured on the web, pre-visit and post-visit windows can be created in the analytical file when building predictive models. The post-visit window, as previously discussed, would contain the behavior to be predicted or the objective function, such as a page view, clicking on an URL, etc. Meanwhile, the pre-visit window would capture all web behavior prior to the post-visit window. With the magnitude of data available, the challenge is to identify or create meaningful variables from this very rich information. As in the offline world, defining the relevant universe for modeling is a critical preliminary step before building any model. Preliminary characteristics that filter out visitors would be used to identify customers clearly not relevant for the modeling exercise because they are very weak performers in terms of the desired modeling behavior. Since online models can be scored in real time, these types of visitors would simply have no score, which means that they represent the worst names. Model scores can be translated to ranks and can thus indicate to the end user the relative degree of model performance compared to the average model performance.

Given the speed of access to digital information, expectations on insights from the data increase. The need for quicker insights and tools requires more robustness in terms of standardizing the analytical environment. With a standardized data environment, analysts can spend more time creating tools and conducting analysis that allow organizations to enhance their ability to glean business intelligence from the data. The data environment under these conditions yields the capacity for developing at least a hundred or so models in a given year simply because there is a standardized analytical file. All potential inputs to any model would be identical for every exercise. The only difference under each modeling scenario would be that the objective function or desired modeled behavior would change. The largest constraint here would be human resources for building these tools and measuring and tracking them. The use of social media and text mining is only going to augment this need for qualified people.

The ability to build individual-level models extends beyond situations where a person has personally identified himself or herself. For example, the general browsing activity of a user visiting a website can be used to identify certain trends or behaviors related to that specific user. Of course, in this scenario, the analytical record of interest is the IP address of the computer rather than the customer. Many leading-edge Internet companies, such as Amazon, Yahoo, and Google use this information and call this "behavioral advertising." For example, based on your clickstream behavior, these companies will place more appropriate ads that better fit your behavior profile. This allows these organizations to become more efficient with their ad placement. Organizations that can deliver superior behavior advertising solutions provide a competitive advantage to businesses seeking to place ads on the Internet. The more efficient a company is in utilizing behavioral advertising models, the lower should be its ad cost per revenue dollar.

After reviewing the different forms of interaction in the digital world, the next issue to discuss, as seen in the offline world, is whether or not the information is available at the individual level. Digital collection of data has provided us with an intensive data environment with individual-level data more readily available. Advanced mathematical tools that were used infrequently by business are now becoming commonplace business practices because of the availability of digital data.

Marketing Attribution

In traditional direct marketing, the ability to directly relate a purchase to a marketing activity is very straightforward. If a publisher is trying to acquire new subscriptions through direct mail, responses can be attributed directly to the marketing promotion either through a business reply card or a specific 1–800 phone number. In digital marketing, this process is somewhat more complex. An e-mail campaign inviting a person to purchase an item at a store does not present an exact linear association between that purchase and the e-mail; after all, the individual might have made the purchase without the e-mail. This is also sometimes seen in direct marketing when the direct mail piece directs the individual to a store in order make the purchase. Yet, historically, traditional direct marketing campaigns have always attempted to tie sales dollars directly to the promotion to the individual. In digital marketing this has changed since e-mail campaigns and social media campaigns may be primarily about awareness; that is, the "ask" is to drive purchase behavior in another channel, such as "going to a store." Yet, digital marketing can operate more directly with specific URLs or 800-numbers in the body of the e-mail leading to a purchase of a given product.

Thanks to the web, marketers now can use multimedia campaigns with digital marketing just one part of the overall campaign. The digital component may be integrated with a direct mail piece, TV campaign, and perhaps an outbound telemarketing campaign. This kind of scenario represents more of the norm rather than the exception in today's marketing environment. The aim of the multimedia approach is to create more awareness and engagement with the consumer that would then lead to a purchase.

In digital marketing, A/B testing can be set up where the A group receives the e-mail and traditional media while the B group receives just the traditional media. Both the A group and B group have the same customer characteristics. In this case, the ROI impact can be tied directly to the e-mail campaign. While dollars cannot be directly attributed to the individual level, the aggregate dollars of each group can be evaluated. The assumption can then be made that any positive incremental or negative incremental dollars in the group receiving the e-mail are due to the campaign. The cost of e-mail campaigns provides the other component that allows marketers to calculate ROI.

The major challenge in web analytics today is to determine precisely whether a click led to an actual sale. If a click leads to a URL that is a call to action, such as the purchase of a product, then there is a direct relationship. But in most cases, as we have seen above, consumers have other options to purchase a given product beyond just clicking on a specific URL. For example, consumers may decide to travel to a store to make the purchase. Yet, the purchase at the store may have been initially driven by some other motivation other than the visit to that web site.

BIG DATA

The discussions and debate surrounding big data will pose new challenges to the data mining community. In the big data vernacular, the three Vs (volume, variety, and velocity) present the key challenges in processing and analyzing data. For direct marketers, large volumes have always been the typical environment when attempting to develop and implement successful marketing programs. However, the notions of variety and velocity of data present new challenges for the experienced direct marketer. Regarding variety, data now arrives in unstructured formats whereas most experienced practitioners are well-versed in the traditional database marketing formats of rows and columns. Yet, before I discuss the newer facets of big data, a discussion of database marketing technology is warranted because this expertise will be applicable to big data. Typically, data arrives in the fashion shown in figure 20.1.

Structured Data			
Customer Nº	**Household Size**	**Postal Code**	**Income**
1	3	L1A3V1	125000
2	2	M5S2G1	30000
3	1	H4B2E5	40000
Transaction Nº	**Date**	**Amount**	**Product Type**
1	JUL 15–2009	100	A
2	OCT 1–2009	75	A
3	SEP15–2009	200	C

Figure 20.1 Example of Structured Data

In this example, we have a customer table and a transaction table. The structured format is determined by the fact that the records (three for the customer table and three for the transaction table) are presented in rows. Meanwhile, the characteristics of the record or variables are presented in columns. Typically, there would be a way to link these two tables, and in fact the customer number on the transaction table here (not shown here for the sake of simplicity) would be the match key in linking these two tables.

Online postings are unstructured in the sense that there are no defined fields or variables. Instead, we have blocks of data that are our raw sources of data to be analyzed. What makes this challenge even more complex is the fact that the data in its unstructured form will also come in different file formats (hence variety as one of the three Vs in big data). The varying formats of data now available require new tools particularly in ETL (extract, transform, and load) tools. In many cases, social media vendors provide APIs to help read the data in these constantly changing file formats. The use of these APIs together with database querying and extraction tools, such as NoSQL, Mongo, Python, etc., allows analysts to extract meaningful information. Another layer of complexity is added by the third V of big data, namely, velocity, which refers to the fact that much of this data is constantly streaming, or constantly "coming at us." From another perspective, this velocity of data simply accelerates the volume of data to be processed.

As new ETL tools are required to handle data in ever-changing unstructured file formats, so new tools are required to handle the data processing requirements and deal with both the volume and velocity of data. Traditional sequential data processing routines are no longer adequate. Techniques involving parallel distributed file processing and MapReduce tools have made organizations such as Apache and its Hadoop technology common names in business today. Without getting into the technical details of what all this really involves, we can say that the purpose of this technology is to process data at significantly higher speeds than is the case with traditional techniques of sequential file data processing.

Integrating these new data processing technologies and the ETL tools allows data miners to operate in the world of big data. However, as has been stated throughout this book, what matters is not data for data's sake but rather data to help solve business challenges. Adopting a "so what" attitude to data helps analysts to understand what data is really required for a given business problem. Is every problem a big data challenge? Of course not, and the real answer is that every problem may potentially represent a data challenge, whether involving small data or big data. In other words, what data do we need to solve the business problem at hand?

This page intentionally left blank

CHAPTER 21

ORGANIZATIONAL CONSIDERATIONS: PEOPLE AND SOFTWARE

THE WORLD TODAY IS EXPERIENCING TREMENDOUS CHANGE IN VIRTUALLY every facet of society: political structures as evidenced by the fall of communism and dictatorships, sports with their new collective bargaining agreements, businesses with their obsessions to become flatter and leaner all with the intention of reducing head count, and technology in its all-encompassing goal of making us more effective with our use of information. However, nowhere is this more clearly demonstrated than in technology and data mining, both of which embrace the use of the best and most relevant technologies to solve a business problem.

Some areas that have changed over the years and will continue to change in the future include:

- Organizational change
- People
- Software

With data and information becoming more and more important as a corporate asset, traditional approaches to corporate structure are becoming less relevant. Technology and increased knowledge empowerment is flattening the organization and ultimately reducing the number of layers between the CEO and employees.

These changes are not unique to corporate structure but are also critical in creating new roles and responsibilities. Knowledge and intelligence and their use for better decision making is the purpose of data and information, and therefore there is a growing need for chief knowledge or data officers. This role is not necessarily an IT role as IT roles typically embrace a deep understanding of technology and systems for the purpose of access to data and information. Using data to derive meaningful knowledge and insights requires different skills. Thinking about data and information in this fashion is vastly different from thinking about how data is best disseminated to users. Knowledge or data officers need to work closely with the CIO (chief information officer), but both roles are equally important if information is the key competitive asset of a given organization. In fact, another potential outcome may be that perhaps the CIO's role evolves over time, and much greater emphasis is placed on data mining and analytics. But for now, let's talk about the role of the chief knowledge officer or chief data officer.

But what are the requirements for chief knowledge officers or chief data officers? For example, in a more centralized approach, all analytical areas report directly to the chief data officer. Thus, in a credit card company, all marketing analysts and credit risk or fraud analysts would report to this person. Although the analytical objectives of each area (marketing and credit) would be very different, the one common theme is that both functional areas use data to derive insights for better decision making. One of the primary roles of the CDO (chief data officer) is to synergize the analytical activities of both areas. This means that analysis should be focused on optimizing marketing behavior while also reducing credit card and fraud risk. The performance objectives of each functional area would be prioritized based on performance within the area, but a significant portion of the performance evaluation would also be based on how the other areas are performing. Through this cross-fertilization of goals and objectives between functional areas, the chief knowledge or data officer can better integrate the activities and tasks between areas so that they are aligned with the overall corporate goals.

DATA MINING IN MARKETING

Historically speaking, no data mining departments existed. As direct marketing evolved into a more common marketing discipline, marketing service departments were created to help execute direct mail programs. Specifically, this involved a number of activities. The first activity was to generate names for a given campaign by working with list broker specialists who would recommend specific lists based on the desired campaign objective. The second activity involved data hygiene and cleansing in order to ensure that the names and addresses were correct and that there were no duplicate names in the promotable file. The third and last activity was generating the campaign list file and its accompanying test cells. Along with targeting the best names for a campaign, the test cells were used to derive specific learning that had been identified up-front as one of the campaign objectives.

From an analytical perspective, minimal activities were conducted in the direct marketing area. Yet, as direct marketing began to grow in prominence, the analytical component continued to evolve and grow due to increased demand for better information and insights. Ultimately, this called for stronger mathematical skills and a certain level of statistical knowledge. With the analytical component delivering and demonstrating significant tangible benefits to the bottom line, both the volume and complexity of work increased. This growing demand ultimately resulted in the creation of either data mining or CRM analytical departments. Typically, in many organizations today, the analytical areas report separately to the functional areas. Examples of this are the marketing analysis and credit risk analysis areas that operate separately. As stated earlier, some organizations are centralizing the analytics or data mining function. The intention of this centralization is to give organizations a more holistic view of customers and ultimately their profitability. This consolidation of data analysis activities will continue to grow as more and more organizations want to create customer-level profitability. However, most organizations still adopt a more decentralized approach toward analytics and data mining with separate analytical or data mining units in their functional areas.

With its tremendous volume of data, the Internet will also reinforce this trend of data analytics consolidation. The key will be to merge online and offline data and, more important, to determine how to use this merged information for better decision making. The world of big data and digital analytics will simply augment the need for a more centralized perspective on analytics and for a chief data officer (CDO). Even without this centralization, the new big data paradigm will enhance the significance of data mining and its role in organizations.

OUTSOURCING IN THE INDUSTRY TODAY

The issue of outsourcing as opposed to insourcing various company tasks continues to be a sensitive topic. The academic, business, and government communities have all contributed many different perspectives on this topic. With increasing technology and globalization, it is simply easier to outsource functions now that were impossible to delegate 10 or 20 years ago. For example, many large organizations outsource their telemarketing functions. Increasing globalization now allows organizations to identify not only where tasks can be achieved most inexpensively but also where the revenues from these tasks can be maximized. Regarding telemarketing activities, many organizations have determined that it is cheaper to conduct these activities in countries with lower labor costs. The cost of training is also reduced as the level of education in these countries continues to improve. Meanwhile, technology has allowed communication to be almost seamless in the sense that one does not know whether a phone call originates in Toronto or New Delhi. These kinds of conditions allow organizations to maximize their profit by selling to the North American market but incurring services or costs that generate this revenue via another low-cost country.

Another example of this is the technology sector itself. Traditionally, technology solutions have always cost money. The cost of a programmer without experience typically ranges from a minimum of $35.00 per hour (including benefits) and perhaps goes to well over $100 depending on the breadth and depth of experience. Low-cost countries in Asia purport to offer these skills at well under the going North American rate. For many companies, certain technology solutions are farmed out to these low-cost providers. However, what is interesting here is that not all solutions are farmed out. Technology solutions that require heavy intellectual capital are conducted in North America, but the more mundane solutions are developed overseas. An example is the design of a new system to meet the needs of business users in a given organization. This design would be done in the North American market, but the coding of screens and visual interfaces might be done overseas.

Evolution of Outsourced Data Mining Analytics

The data mining industry is certainly not immune to the debate concerning outsourced activities. As with many other services, the need for data mining developed in certain organizations that found it a critical component of their business. Even 15 or 20 years ago, only a handful of companies attempted to do this kind of work. These few organizations invested heavily in technology and resources that were required at that time to engage in data mining activities. Many people working for these organizations realized the tremendous business value of data mining and saw that technology was the limiting barrier at that particular time. This situation changed in the 1990s when technology barriers were eliminated, which allowed many more organizations to conduct data mining activities as part of their core service deliverables.

As organizations are more empowered in data mining, they need to determine whether activities should be outsourced in full, in part, or not at all. The solution will depend on the type of organization and, more important, on the key stakeholders and decision makers.

Core versus Non-Core Activities

At an organizational level, company activities and tasks are segmented into core and non-core competencies. For companies considering data mining a non-core activity, the solution is simple. In almost all cases, much of the work will be outsourced but managed internally. Some of the more mundane data mining activities, such as basic reports, may be done in-house depending on the data and analytical skills of the company's personnel.

Meanwhile, companies deeming data mining a core competency will be more reluctant to outsource mundane and the more advanced activities. These core competencies are often considered to be strategic in nature and therefore a critical competitive advantage of the company. For the most part, this would include banks and telecommunications companies. In determining whether data mining is a core business activity, the following questions should be asked:

- Does data mining provide critical intelligence and business advantage in the selling of communication devices and services for telecommunications companies and financial products/services for the banks?
- Can data mining provide strategic insights that are key competitive advantages both for banks and the telecommunications sector?

For these above-mentioned types of organizations, the answer is yes. However, whether or not data mining is a core competency that must be retained in-house is still up for discussion. In determining the answer, it is important to understand whether or not data mining solutions are transferable between organizations.

Do Core Data Mining Activities Need to be Solely Done In-House?

For larger institutions, such as banks and telecommunications companies, the strategic importance of data mining and its identification as a core activity will be obvious. However, data mining's many deliverables are often tactical in nature. For instance, developing models to target names is a tactical solution to a specific business need. Certainly, models are critical components in providing services and products to an organization's customers. But does this work always need to be done in-house? In other words, does the company lose an advantage by outsourcing this function to another company? Certainly, the outsourced company will have information pertaining to customer characteristics that are most pertinent for a given customer behavioral model. But this information is unique and specific to that organization, and this learning is not easily transferable to other organizations. This is the crux of the issue. If the learning from a data mining exercise is nontransferable, then it does not matter whether solutions are developed in-house or not.

Solutions themselves are not really transferable between organizations. In regard to data miners' knowledge, should this remain in-house or outsourced? One view is that companies best utilize the knowledge base of data mining for their own specific business needs. This suggests that the goal of keeping the knowledge base in-house as opposed to using external resources is not the real issue. The important considerations here are where to obtain the best value and services and how to do so inexpensively. From an organizational standpoint, executives should ask the following questions when deciding to outsource data mining:

1. Can I meet all the demands for this work in-house and at lower cost?
2. Do I have the required expertise to do this work optimally?
3. Does a different point of view and perspective yield insights that can further optimize business decisions?
4. Does a complementation of in-house resources and outside expertise deliver incremental value?

The discussion so far has dealt with outsourcing purely from a business rationale. However, more subtle and hidden reasons can impact the decision to outsource. These reasons are often more psychological and can be attributed to people's reactions to security. All of us desire to be secure in whatever environment we are in. Certainly, job security is one specific example of this. Here is an example of how this psychology might work. Let's suppose a person, John Smith, runs a marketing services department of 10 people. John is told that outsourcing must be considered because there is more demand for services, but these services must be completed with fewer people in-house. His first reaction is the realization that the value of the department to the organization has been diminished due to these outsourcing considerations. Moreover, he is expected to do the work with fewer people. But rather than viewing this issue strictly in terms of headcount, he could see it from a budgeting perspective. In other words, the significance of his department relates directly to the budget that is allocated to his area and not necessarily to headcount. However, he must determine how to influence the dollars allocated to his budget; that is, he must do two things:

- Demonstrate the benefits of data mining activities and their impact in the organization
- Ensure that the senior people in the organization clearly understand these benefits

Merely increasing the headcount in the department will not accomplish this. Yet, if John is focused on maximizing the benefits of data mining in the organization, then he will try to find the best resources both in-house and externally to achieve the objective. If John can demonstrate benefits to the organization that result in $15,000,000 being allocated to his area, it does not matter whether he can achieve these benefits with 15 people or 1 person. The priority is to get it all done in the given time frame and to obtain the best solution using both in-house and external staff. The adoption of company policies that exclude outsourcing as an option can be very limiting. Even with highly talented in-house staff, external suppliers, given their broad range of industry experience, can bring a point of view that can yield unique insights for a given business problem.

An important component in addressing job security is not to build up an empire in terms of headcount but rather to build up the required budget for this area. Budget increases are essential to job security. But the key to maintaining budget amounts or even to obtain budget increases is to continually achieve the benefits of data mining, and this is best accomplished through a combination of both in-house and external resources.

Another growing trend in the area of outsourced analytics is conducting advanced analytics offshore. As discussed previously in the telemarketing example, offshore firms now conduct analytics that were traditionally the sole responsibility of the onshore firm. These activities tend to be more technical in nature and require heavy programming together with the application of advanced mathematical techniques. Meanwhile, the onshore company has the depth and breadth of data mining experience across a wide range of industry verticals and can thus manage the process more efficiently by not conducting all the requisite data mining tasks. In practice, however, there have been mixed results regarding the efficiency of this process in terms of streamlining costs for a given data mining project. The question organizations must answer is whether it makes sense to hire junior data mining analysts who initially would execute much of the same activities as the offshore organizations. In-house training and mentoring of these junior analysts might be considered an investment as the knowledge and experience of these junior analysts would be superior to what might be obtained by offshore analysts working with onshore companies. Onshore organizations must never stop investing in new analysts from their own shores as the potential payback is great. The deeper problems to be solved through data mining can only be resolved by individuals steeped in academic training as well as practical experience obtained though mentoring and job training. At the same time, offshore companies must be considered viable options if the technical tasks follow clear and straightforward processes.

PEOPLE FACTOR

The people factor has always represented the greatest challenge in building a data mining practice. Since the data mining discipline is relatively new, finding knowledgeable practitioners has been difficult. In the past, organizations relied on finding individuals from other organizations with an extensive direct marketing discipline. Formal educational training in this area was nonexistent.

However, this has changed as educational institutions, such as Dalhousie, and Queen's University, now offer business degrees that focus on data mining. One common component in these educational programs is how data mining is deployed in marketing. Community colleges also offer specific courses in data mining, and business associations, such as the Canadian Marketing Association and Predictive Analytics World, have two-day seminars and courses devoted solely to data mining and predictive analytics.

Despite these educational developments, formal education in data mining is still in its infancy. Where people are concerned, though, changes will occur through education, specifically at the university level. Besides Dalhousie, Queen's University now offers a master's degree in business analytics. With the community colleges also embracing this discipline, it is only a matter of time before more universities become more engaged in education in this sector.

Changes will happen in the universities themselves as the computer science departments evolve into specialized disciplines, such as data mining or perhaps into a more all-encompassing area called knowledge management. With or without these changes, though, data mining will still become a core component in any information management discipline. This increased level of training should equip young individuals for junior-level positions in data mining. Organizations can then provide the much needed practical training to complement this academic training and create more knowledgeable employees.

Software Changes

The software component of data mining represents the one area of data mining that has seen the most change. This will continue in the foreseeable future. Why has the software world changed so dramatically? With cheap and easy access to data through technology, there has been a huge demand to empower more people to do data analysis. Historically, this capability was limited to highly technical people with strong programming skills. In particular, those with SAS (Statistical Analysis System) programming skills were treated as company's data mining gurus. These skills are still highly valued in companies, but there has been a shift to empower other individuals. First, regular business analysts have been empowered to do nonstatistical data analysis. Second, organizations empower more mathematically oriented individuals in conducting statistical analyses; however, these people's capabilities are limited because they do not have strong programming skills. In these two examples, new software can help facilitate data mining by eliminating the need for someone to write programming code. Instead, graphical user interface (GUI) provides allows analysts to conduct a data mining exercise. The analysts still must have a deep understanding of the data mining process, but they can conduct this exercise without writing any programming code.

Another significant improvement in data mining software is the development of tools that conduct real-time analysis. In other words, analyses can be conducted any time with the most up-to-date, accurate information. This can have tremendous relevance for online marketing where information is constantly being exchanged. Having tools that can access data and conduct analysis at any time represents a very significant improvement when compared to the past practice of having to wait for the next mainframe database update or wait six to eight weeks before having enough data to analyze a direct mail campaign.

Many of these new tools now have high-powered mathematical components to help target specific prospects and/or customers. Although it is exciting for data miners to have access to these new high-powered tools, these are just tools and no substitute for skills, knowledge, and experience.

Software vendors and high-powered mathematicians might say that their new tool is the next breakthrough in mathematics; however, data miners must be "grounded" in evaluating these tools. Proper validation exercises are critical for effective evaluation of these tools.

Perhaps the most exciting developments in data mining are coming from text mining. The learning and work conducted in this area include techniques that focus on processes to convert unstructured data, such as text, into actionable structured data. This is done through a process involving data parsing/cleansing, and conversion of the data into actionable numeric data, and classification of this numeric data into categories using certain clustering techniques. Consider the opportunity here. Data miners can now use any form of data, written comments, and other text data on customers to uncover certain customer patterns that might be relevant for a given customer behavior. An example of this is the ability to examine data from e-mails between customers and a retail company. With customers' e-mail data text mining analysis may indicate that texts should be classified into three categories, such as complaints, request for catalog information, or request for specific product information. This data could then be used to determine whether any of these three categories have an impact on retention. The ability to identify and discover patterns in this type of text data is gaining more and more prominence, especially in big data.

As the data mining industry grows, products are being developed that empower more business users in this discipline. These tools allow high-end business users (statisticians, programmers, mathematicians) and low-end business users (business analysts) to conduct data mining tasks more efficiently. High-end users can use sophisticated statistical analyses very quickly. They can try a number of different mathematical approaches and observe results. The software allows analysts to spend more time in analytical thought, which will allow analysts to use their interpretative skills in evaluating results from a number of different approaches. For example, five different techniques might be utilized: logistic regression, CHAID, neural nets, genetic algorithms, and linear regression.

At the same time, let's suppose a data environment is to be created that does the following:

- Create index variables on all continuous variables
- Create change variables on all variables that have both a time and value component
- Create categorical variables using CHAID
- Create categorical variables using factor analysis
- Create categorical variables using cluster analysis

The software tackles these challenges by first providing all the mathematical techniques needed. Analysts then must interpret these results and assess what the impact is for the business. In addition to conducting high-end mathematical routines, analysts should also be able to create new derived variables; as discussed previously, this is an extremely important data mining skill. At the same time, software is only a tool, and the identification and creation of derived variables relies on the insight and knowledge of data mining practitioners.

From a more general or philosophical perspective, the guiding mission behind data mining software is automation—not automating data mining itself but rather automating tasks that require minimal critical thought. Instead, software advances now allow analysts to devote more critical thought to the problem and to consider different approaches and techniques that might help solve the problem. The mundane task of creating programs to generate the necessary information for critical thinking is now automated. An example of this is the development and evaluation of models. Once the analytical file is created, multiple models can be very quickly created without requiring programmer resources. Using a GUI, the analyst can select the appropriate technique and can develop a model on the fly. Gains charts or decile tables are automatically produced to evaluate the results of the new model. This automated development of models allows analysts to look at many different options. Analysts now spend more time considering different modeling options rather than in writing programming code. They still must understand in detail the components or variables of the model because data miners are responsible for knowing the details of any black box solution. Model results of any business initiative must also be validated. How does this occur effectively in an automated modeling environment where hundreds of variables are created? Automated routines now look at the KS (Kolmogorov-Smirnoff) statistic or the volume of that area of the curve between the flat line and the parabola (lift curve) that has been discussed previously. An example we saw previously is in figure13.4.

Ranked Intervals

If the change between the KS statistic (area under the curve between the parabola and the straight line) in model development and its current application fails to reach a certain threshold as determined from model development, then programming routines can be written to tell users that the model has significantly eroded. Organizations such as SAS offer specific tools in this area, such as the SAS Model Manager to help identify when these triggers occur.

Evaluating Software

The first and primary considerations are the actual software solution's results. Are better results produced with other software options or with existing software? As we saw above, looking at the KS statistics of different software solutions is one option for evaluating a software package. Another option is to compare how a given solution rank-orders the results, and this is another way of viewing the KS diagnostic. If the primary data mining objective is to differentiate results, then comparison of results across different software solutions might yield the graph shown in figure 21.1.

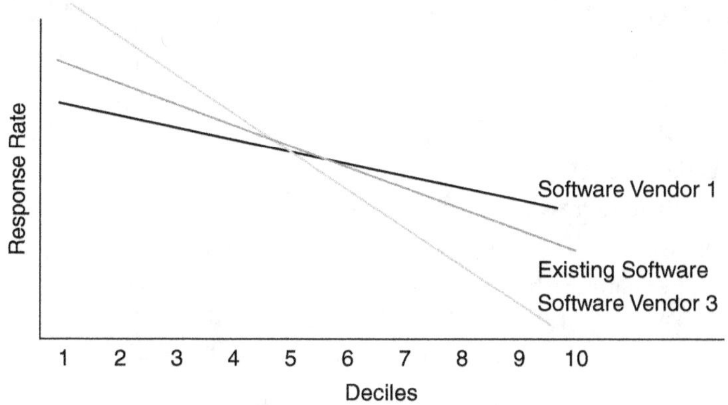

Figure 21.1 Evaluating Different Software Solutions Using Lorenz curves

In our above example, software vendor 3 would be the optimal choice because it provides the best rank-ordering results from the top decile (decile 1) to the lowest decile (decile 10).

Another factor in evaluating software is its ability to create derived variables. High-end users are programmers who can manipulate source data from a number of files and then create new variables. Being able to create derived variables with minimal programming saves programming time but also empowers other users who may not have a programming background. In this case they would use the appropriate interface and modules as pointers to do the following:

- Identify source files and fields that will be used in the exercise
- Link appropriate files and use appropriate match keys in linking these files
- Run mathematical and statistical routines to summarize and manipulate data into other variables

Even though software may facilitate the above-mentioned functions, most advanced practitioners currently still rely on their programming and technical skills for these tasks. Yet, as data mining continues to evolve, these less programming-intense tools will be utilized more fully by the data mining community.

In addition to being able to derive new variables, the software should also have flexibility in terms of including a variety of statistical solutions.

Both linear and nonlinear solutions should be examined with such techniques as:

- RFM
- Linear regression
- Logistic regression
- Neural nets
- Genetic algorithms
- CHAID/CART

Scalability as a Consideration

Scalability is an important consideration since expanding organizations will presumably have expanding data requirements. This is important for big data. Questions should be asked as to whether the software has specific limitations regarding the amount of data that can be included in any application. Users should also be aware that certain mathematical routines are much more sensitive to increased volumes of data.

Portability of Output as a Consideration

Often enough, the output of certain applications must be input into other applications, such as Excel, Word, or PowerPoint. How easy is it to output data to more user-friendly business applications? In the early days of data mining, this task was quite onerous and often required much cutting and pasting as well as typing of raw output from a statistical application into a more user-friendly application. In today's environment, raw output can be sent directly to an Excel file. This has resulted in significant time savings, and data miners can now spend more time on analysis rather than having to massage raw output for presentation purposes. Besides the portability of output to other applications, the inputs into the application must also be considered. For example, how compatible is the software for receiving data from a variety of different formats, such as comma-delimited, tab-delimited, SPSS datasets, SAS datasets, etc.?

Reports and Diagnostics as a Consideration

The ability to produce the right reports and statistical diagnostics is important in producing the optimal solution. This is best achieved by software applications that have flexible reporting capabilities that are easily generated by end users. In addition to these reports, there must also be a functionality that creates standard reports with minimal user intervention.

The use of statistics in any applications should provide sufficient diagnostics to allow users to mathematically compare different solutions. For example, the use of R^2 in linear regression or concordant/discordant pair ratios in logistic regression represents specific metrics for comparing different statistical solutions. Yet, the most important reports in these software applications yield information on the business impact of a given solution. In building models or targeted solutions, this means generating gains tables or decile tables, which have been previously discussed at great length.

Gains charts, as seen previously, demonstrate how well the response model applied to a given campaign differentiates segments (deciles) of customers based on response rate. At the same time, revenue numbers and cost numbers can be used to convert the response rate performance into an ROI metric. The ROI numbers can then be used to select customers that would be most profitable in a given campaign.

Looking at Software from the Business Analyst's Perspective

Recent developments in software now allow analysts to conduct their own analysis. These analyses do not require the intervention of high-end users such as statisticians or data miners. Instead, analysts can obtain top-line results of a previous campaign or business initiative. Generating results for various campaigns or conducting various ad hoc analyses does not require sophisticated analysis. Instead, users create cross-tab reports that yield the necessary information and insights on a particular business problem. Users do not need to write programming code to generate reports but simply use the GUI (graphical user interface) to generate the appropriate reports. However, users must understand the mechanics of the data, such as what information is to be measured in terms of performance and how they want to view these performance measures. Examples of this might be a report on income (performance measure) and creating views of this report based on gender and region.

This kind of empowerment means purchasing software, namely, business intelligence or BI software. The technology behind this software is based on creating data cubes. These cubes in theory represent prebuilt queries of the data, and the design of these cubes addresses all possible queries of the data.

In assessing this type of software, the primary consideration should be ease of use. In other words, how quickly can analysts learn to use the software? Ease of use can translate into more users having access to data. This enhanced empowerment of end users means that more analysts are dealing directly with the data rather than having to go through a programmer to generate reports. Ultimately, this will lead to better decisions due to quicker turnaround of results and information.

In addition to ease of use, the other key consideration is the range of variables that can be used in designing reports. Obviously, a more extensive range of variables will provide enhanced flexibility in generating reports.

In choosing specific software, considerations will vary in importance according to an organization's needs. For instance, an organization beginning with analytics may look at simple tools, such as BI software, that tackle only a limited number of variables but offer ease of use. Organizations more advanced in analytics may place less emphasis on ease of use and more on the variety of statistical tools available in the software. As with any purchase decision, cost will always be an important factor. Its weight, though, in the overall purchase decision will depend on the importance of other considerations in the decision-making process.

Choosing software for an organization is not an easy process but requires a deep understanding of the organization's analytical needs. With an understanding of these needs, analysts can give the appropriate priority to the key considerations as they select the software that addresses their organization's specific analytical needs.

CHAPTER 22

SOCIAL MEDIA ANALYTICS

PUTTING A DIFFERENT SPIN ON SHAKESPEARE'S FAMOUS PHRASE IN *Hamlet* "To be or not to be" captures the question of analysis and what it means in the social media sphere. Because of the nature of the medium, privacy implications, consumer sensitivities, and perceptions become more complicated. What do I mean by this? Traditionally, privacy concerns were focused on the fact that a consumer has had some interaction with an organization. The information concerning this interaction is captured and can be used for marketing purposes. The laws around PIPEDA (Personal Information Protection concerning Electronic Documentation Access) are quite clear about how marketers can use information regarding the transfer of information between the consumer and a given organization.

But the world of social media raises the debate to a whole new level. Consumers are now interacting and engaging with a variety of individuals and organizations in their community. For example, I may be a valued CIBC customer who has responded to an ad on Facebook regarding some RRSP promotion. This information pertaining to my response would certainly be captured by Facebook. Now, Facebook would have all the other data pertaining to my interactions in my social network. For instance, my plans for Christmas, which I would certainly discuss in my forum of friends, might be of extreme value to a CIBC marketer in promoting certain products to me at that time of year. Remember, I am a customer of CIBC, and the company is merely attempting to serve me in the best manner possible. Yet, the information the company may be using in determining the right marketing approach to me may be generated from information pertaining to my Facebook interactions. Currently, this information is subject to controls in the sense that Facebook owns the data and would probably look unfavorably on handing over personal information regarding its clientele to a major bank. More important, it is highly unlikely that a bank or other leading ethically oriented business organizations would violate customers' trust by using information customers expect to be restricted to their social network. However, the laws are not necessarily clear and certainly do not stipulate how organizations should use information in social media.

Within the Twitter space, the privacy issues can even become more complicated, as the data is freely accessible to anyone including organizations. In the current environment, organizations potentially have access to the data of all their customers who interact on Twitter. The key, of course, would be to get customers to engage in some type of company-initiated promotion. Presumably, if a current customer of an organization responds to a company promotion, the customer would in effect reveal his or her customer identity within the organization. At the same time, the customer's log-on identification could also be attached to his or her record, and this in effect allows the organization to gather that individual's behavior on Twitter. Clearly, this kind of data environment and lack of any real policing when it comes to privacy protection opens up the door to all kinds of abuses. There are no clear-cut answers in this area, but practitioners need to be aware of these issues because the data is available for their use.

If these privacy concerns can be addressed, the ability of organizations to gain information about the social media behaviors of their customers can simply lead to more effective marketing. As with all data, the most significant benefits arise from the analysis of all behavior of individuals as well as from acting on the resulting learning on an individual level.

From a targeting standpoint, additional variables can be created and included in any predictive model. In addition to the targeting benefit, social media provides unique insights on how best to communicate with customers of a given organization. For example, one key objective of analyzing any social network is to identify the influencers and distinguish them from the followers by analyzing the network activity of each record. Activities such as the number of contacts or friends within individuals' network, the number of times they reach out directly to their friends as well as the number of times that friends reach out to them along with other types of activities can all be used to identify influencers and followers. For example, through network analysis the contacts that an individual has made in a given period of time can be analyzed, and by examining the breadth and depth of the individual's network, the company can identify whether that individual is an influencer or a follower. Breadth refers to the number of contacts in the person's network, and depth refers to the level of engagement each person has with his or her contacts. Various metrics and indices can be calculated that allow the analyst to create an "influencer score." A component of this type of analytics exercise would be to determine the threshold from these metrics/indices that separates influencers from followers.

Once we have separated customers into influencers and followers based on their social media behaviors, we could formulate marketing communication strategies that lead to different types of messages to those two groups. Using the same approach in creating RFM in the more traditional CRM programs, marketers can define similar key social media behaviors. For example, habits related to average length of time for a given session, most common time of day when logged on, and how often individuals log on in a given period of time can all be used to derive insights regarding consumer purchase behavior. These pseudo-RFM behaviors can be examined for both influencers and followers. These behaviors can then be further analyzed by exploring the different user groups the individuals belong to.

Another challenge in social media analytics is the issue of marketing attribution, the determination of how much ROI from a given marketing campaign can be attributed to social media. There may be no right answer to this challenge. Unless the social media campaign includes a direct method whereby a given person logs onto a certain web page and registers for a given product or service, it is very difficult to determine the dollar amount of a sale that can be attributed to social media. Yet, even when using this functionality, one could argue that nonregistrants participating in a Facebook contest might be contributing to ROI by purchasing at a store, through TV, or through direct mail as a result of just opening up the Facebook contest page. The same challenges of attributing ROI to an advertising channel exist throughout the mass advertising industry; however, one key advantage of social media is that the denominator in determining the number of people who clicked on a fan page or contest is known. Because in most cases we cannot directly attribute a social media sale to a specific individual, analysts can only provide general information on whether a campaign created additional engagement and whether or not this additional engagement translated into additional dollars.

On the engagement side, metrics, such as the number of people logging onto a fan page or contest and how long they stay on the site, can represent some kind of engagement proxy, and these metrics can be compared with those of other social media campaigns to determine the level of success. By the same token, anecdotal analysis of social media campaigns over time can determine the overall impact on incremental sales. Classification of campaign periods into high, medium, and low social media marketing allows executives to see the general impact of social media marketing on sales. However, one cannot directly trace the ROI to a specific social media campaign. All that can be concluded is that the campaign generated X % improvement in engagement relative to other campaigns. The leap of faith for executives is that X % improvement in engagement translates to incremental dollars, but the precise number is unknown.

Assuming that privacy concerns are addressed, the more granular types of analysis, such as capturing social media behavior at the individual level, will always be the preferred option of marketers. Of course, this does not mean that meaningful analysis cannot be done unless we are able to overcome privacy restrictions.

All the records in a social media network can be analyzed without knowing the name and address associated with a specific record. For example, as outlined above, influencers and followers can be identified even without knowing the identity of the individual record. The social networking and online habits of influencers and followers can be examined, and records can be separated into the different user groups in the medium. The difference here is that targeted marketing to individual online users cannot occur. Yet, specific communication strategies can be developed from insights drawn based on the behavior of different user groups, such as influencers and followers. Rather than deliver one message for all users, specific targeted messages can be created for a specific social media site and potentially for different social media user groups.

As social media marketing develops, the analytical possibilities will expand, but privacy concerns must be addressed. In addition to better understanding of how social marketing fits into a given organization's marketing strategy, marketers must gain a better grasp of analytics and how it fits into a given social marketing strategy.

Chapter 23

Credit Cards and Risk

Fair Isaac

The birth of credit cards in the late 1940s and 50s led to the demand for techniques and processes to reduce overall credit losses.

The bulk of this responsibility rested on the credit operation and collections area of these organizations. Business rules were the norm; that is, if a person's credit reached a certain level, various procedures were introduced to reduce the likelihood of the debtor going into default. Examples of these procedures could range from a phone call indicating the organization's concern with a client's current credit risk status to actually shutting down the account until payment was received. These procedures were successful in reducing overall credit losses. However, given the amount of these losses on the credit companies' books (in many cases it was estimated to be in the tens of millions dollars), senior company executives were always considering other techniques and approaches to reduce these losses even at a marginal level. An improvement of 1% on a book of $50 million losses translates to $500K in annual direct savings. With the ability to reduce losses at this scale in perspective, companies began to consider other options or ideas that would reduce their overall risk exposure.

One business tool under consideration was the use of statistical techniques (namely, multiple regressions) as way of targeting high-risk customers. During World War II, statistics were first used to predict certain enemy behaviors and tactics based on prior history. The success of these measures led business executives to consider statistics as a potential business tool.

The need for these types of tools led to the formation of a company called Fair Isaac Corporation. Using mathematical techniques, such as multiple regression, the company created a scoring system that created risk scores based on an individual's prior credit behavior. In this system, individuals who applied for a credit card were told that their credit history would be used in determining whether or not they would be approved for a credit card. At the same time, individuals were told that if they were approved, their ongoing credit information would be assessed periodically.

For new applicants, the process of determining approval involved comparing an individual's score against a selected cutoff range. Individuals who scored above the cutoff were accepted for the card, but those who scored below that were declined.

On an ongoing basis, credit card customer-related information was sent to Fair Isaac where updated scores based on an individual's current credit history were attached to the customer's credit card record. These scores were not meant to replace the current credit business rules and procedures for high-risk customers but rather to augment them. With the scores, the credit card companies had additional information that ultimately increased their decision-making capabilities. These developed skills led to a refinement of the current credit business rules and procedures concerning the treatment of high-risk customers. Like any sound organization, credit card companies compared these new rules and procedures to the status quo to determine the overall net impact on the business. After the evaluation, the new rules were implemented, and they proved successful as Fair Isaac is still in business today and is regarded as the preeminent expert company when it comes to modeling credit risk. In fact, the power of their capabilities is best exemplified by the longevity of their models. In many cases, the models and their scores have been in use for 5 years without any major updates. In some cases, the longevity has gone as long as 10 years.

The success of Fair Isaac with these techniques led other organizations to become more proficient in this area themselves. Recognizing the huge value of loss prevention even in a marginal amount, organizations hired their own statisticians and mathematicians so that some mathematical expertise could be devoted to their business. Organizations now produce their own tools in conjunction with Fair Isaac's solutions to further enhance the decision-making capabilities of their credit operations.

In Canada, comparable organizations, such as Equifax and TransUnion, offer credit bureau services as well as mathematical services based on scores. The Equifax Beacon score is a commonly accepted credit risk score used by companies to assess a customer's creditworthiness. In addition to the scoring of customer records, these companies have collected credit data regarding customers and have created huge customer databases. This lucrative activity allows these organizations to conduct what is called a credit report on individual customers. Given that the customer has consented to this, any financial institution offering a loan product can obtain a credit report on a potential applicant. This credit report is essentially a copy of the customer's record detailing the person's current credit history. Using key information and fields from the customer's record, companies can use a combination of credit score and credit history information to make their credit approval decisions. An example of this might be that credit card applicants must have a credit score above 650 and have only once in the past 12 months gone 60 days without making a payment.

In fact, the success of using credit scores in assessing the likelihood of credit default has crossed into the area of B2B (business-to-business marketing). Companies such as Dun and Bradstreet have offered such services for years. Paydex scores (a measure of the company's likelihood of going into payment default) is used as a key metric by most organizations when they decide whether or not to extend credit to other organizations. In addition to the Paydex score, Dun and Bradstreet also collect a vast array of information on payment history, which is transformed into indexes by industry sector. In addition to payment history, which is collected and reported at an industry level, information pertaining to the demographics of a company is also available, such as company size, tenure, sales, industry sector, etc. Organizations can use this information in their business decision-making processes and to potentially develop models that target response as opposed to risk.

CREDIT RISK AND MARKETING BEHAVIOR

Although the pioneering techniques of data mining were developed in the credit industry, the extension of these techniques to the marketing realm of the business world has also resulted in marketers using credit risk information as key component in their own decision-making processes. Credit risk behavior has been shown to be a powerful predictor of consumer behavior. From a marketing standpoint, this makes sense in that a particular individual's credit patterns will influence that person's decision to purchase a particular product or service. The best example of this behavior is the credit card industry. In building models to predict response or the likelihood of someone filling out a credit card application, it was discovered that the most responsive prospects were individuals who were also more likely to seek more credit. These credit seekers were indeed high-risk credit customers. This information was confirmed when the initial response models performed very well in getting prospects to fill out credit card applications. However, average approval rates (which involved checking the credit report of these prospects and then deciding to reject or approve them) decreased dramatically with the introduction of these models. The next step was to build models that optimized both the likelihood of customers filling out an application as well as the likelihood that they would be approved.

Aside from the credit card example, there have been other cases of strong relationships between marketing and credit risk. In the auto and property insurance industry, the likelihood to purchase policies is positively affected by good credit behavior. From the perspective of claims risk, higher credit -risk is associated with increased likelihood of having a claim. Meanwhile, the retention behavior of insurance customers generally seems to indicate that customers with higher credit risk are more likely to defect.

TRIGGER-BASED BEHAVIOR AND FRAUD RISK

In the world of data mining, one of the new terms often mentioned as a key concept for marketers is that of trigger-based marketing. Trigger-based marketing is the identification of key marketing behavior such as purchase behavior that is out of pattern. For instance, a customer typically spends $100 per month, but in the past three months, this person has been spending $500 per month. Clearly, the customer's behavior has changed. This type of information is invaluable to marketers because developing products and services in relation to this change will always be a marketing priority. Examples of trigger-based marketing can be found throughout history. More specifically, life-stage marketing is a form of trigger-based marketing.

Specific services and products are required depending on the various stages of life and situations of consumers. Since most companies have hundreds of thousands of customers, the challenge is individually identifying the particular life stage and situation of each customer. The use of a marketing database can facilitate this process, but it will be less than perfect. For example, key information, such as age, gender, number of household occupants, education, income, occupation, etc., may or may not be contained in a database. Even if these fields are contained in a database, it is highly likely that many of these fields will have missing values.

Determining the precise life stage of a particular consumer will depend on the data. Given the data environment, broader assumptions regarding a consumer's life stage can be inferred by marketers. For example, marketers may look at a group of people over 60 years of age and specifically consider the group's spending patterns with regard to a company's services and products. They may find that the customers in this group have decreased their spending by over 50% in the past two years but are maintaining this lower level of spending. The significance of this finding is that there has been a major decline in spending, but this decline represented a one-time effect rather than a continuous decline in purchase behavior, which ultimately leads to outright defection. Marketers may infer that the people in this group aged over 60 who decrease their spending but maintain it at this new lower level are retirees. Although this might not be the exact situation for the entire group of customers, marketers could design a communication strategy that recommends specific retirement products and services without directly identifying the consumer as a retiree.

Another example might be a customer who is married and who ultimately increases spending. Further analysis of the spending pattern may indicate that much of the new spending is attributed to products related to kids and infants. Once again, this type of knowledge can be used to develop specific strategies related to this change in life stage.

Although the development and execution of trigger-based marketing programs is a relatively new phenomenon, the notion of identifying and determining segments and groups most impacted by trigger-based marketing is not new. For example, fraud models have been around since the inception of credit cards. Businesses recognized the tremendous benefit of detecting fraud and ultimately preventing it. The ability to do this was based on being able to detect irregular spending habits; and to then create a flag on the account at the point of sale. This means that when the card is used, a flag notifies the merchant at the point of sale that further investigation is warranted before accepting the sale. Further investigation might include a call to the credit card company to review the situation. But an important aspect of creating this fraud flag is the ability to detect the trigger or out-of-pattern spending. This trigger will vary from company to company; yet, in principle, the goal is to identify actions that differ significantly from the norm. The use of simple variance analysis might be one approach for examining behaviors at different numbers of standard deviations from the norm. Validation of this approach would be assessing how many frauds are identified with the current set of rules. By employing

# of standard deviations from norm	% of all actual frauds captured in sample	Fraud Rate
0.5	90%	45%
1	80%	70%
2	50%	75%
3	25%	100%

Figure 23.1 Example of Fraud Capture at Different Standard Deviations from the Norm

some sensitivity analysis, the percentage of all actual frauds captured can be identified using different sets of triggers (number of standard deviations from the norm). Meanwhile the fraud rate is the number of actual frauds divided by the number of observations in the sample that are outside the threshold of the number of standard deviations. An example is provided in figure 23.1, and here a good cutoff rule for identifying an appropriate trigger might be 1 standard deviation. Below one standard deviation, marginal improvement occurs (increase of only 10% of all frauds); yet, a decrease in the fraud rate of 25% is observed.

In addition to using variance analysis, which might be the quick and easy approach, analysts could use the more robust predictive modeling approach. This approach identifies records that represent fraud, as opposed to those that do not, in a particular time period and then identifies the particular demographics and behaviors of those records prior to this period (pre-period). Constructing the information in this manner allows analysts to identify those key behaviors and demographics that predict the likelihood of fraud occurring in some defined period. The determination of these periods (pre-fraud and post-fraud) is critical to the success of these models. For instance, the post-fraud period probably represents the hour when the fraud was reported by the customer to the company. The pre-fraud period in this case would be very short, perhaps consisting of the past month of spending prior to the fraud being reported. For events that are not fraud, the pre-period would be the current month of activity. From an intuitive standpoint, it is known that the critical fraud behavior occurs just hours prior to its detection. The question thus is why the benchmark isn't several hours prior to the event. The problem is that credit card spending by nature can be very volatile due to the various merchants and establishments that accept the card. There needs to be an established time horizon that can smooth out this volatility. Therefore,

% of Customers Ranked by Fraud Model Score	Minimum Fraud Score in Interval	Average Fraud Rate
0–10%	0.051	4.40%
10–20%	0.045	4%
20–30%	0.042	2.90%
30–40%	0.038	2.70%
...		
90–100%	0.0045	0.50%

Figure 23.2 Decile Chart of Fraud Model

the behavior in question needs to be benchmarked for a time window that is a reasonable representation of that person's purchase behavior. In other words, is it rational to expect customers to use their card once per month? Or do they use their card more or less often?

Typically, good customers can be expected to use their card every month and accordingly based on that learning, one month can be used as the spending benchmark for fraud detection. The actual threshold for determining whether or not a record is fraudulent would be based on a score that yields a fraudulent rate that is unacceptable to the organization. In the example in figure 23.2, this threshold might be all records in the top 20% as representing those that would be considered fraudulent.

Once again, as with any approach, the solution and any resultant decision making can be validated based on how the method performs.

This page intentionally left blank

CHAPTER 24

DATA MINING IN RETAIL

HISTORY OF THE RETAILER

When considering consumers and their behavior at the most basic level, typically the transaction between seller and buyer comes to mind first. Being able to sell what a consumer wants is at the core consumer marketing. In earlier times, buyers and sellers converged at a public venue in open forums. Transactions then involved bartering, and buyers and sellers exchanged roles during the transaction. Eventually, goods and services came to be exchanged with some form of common currency rather than through bartering. Currency as the primary exchange instrument for buying and selling made more formal structures possible, such as stores.

An early type of store was the mom-and-pop store; these stores served a local community with a limited population and selection of goods. Current equivalents of the mom-and-pop stores are convenience stores such as Becker's and 7/11. In these stores, sellers often know their customers personally. The song lyric "Where everybody knows your name," associated with the TV program *Cheers*, represents the retail philosophy of these stores. In these mom-and-pop stores, we can see the CRM philosophy at work: one-on-one marketing prevails.

With the population shift toward urban and suburban areas, the mom-and-pop stores have developed into the larger retail organizations we are all familiar with. In this environment, one-on-one marketing or selling is no longer viable. Instead, mass marketing now promotes products and services to an ever-expanding population. However, this mass marketing approach, despite its continued popularity in retailing, does have its limitations. Retailers are facing increasing competitive pressures such as price, consumer access to information, loyalty programs, and, perhaps even more important, less patience of consumers to hear the retailer's message given all the other demands on consumers' time.

RETAILING TODAY

How have retailers adapted to this new paradigm? Realizing that data and its analysis are key to better understanding customers and ultimately better serving them, retailers adopted processes and procedures to achieve their objectives. Retailers face unique challenges when compared to banks or insurance companies. Banking and insurance transactions are gathered electronically and at the level of the individual customer. The data or information is typically captured with an account ID that can then be generalized to the customer level. For example, a client's deposit, withdrawal, borrowing, investment, and credit card behavior can all be viewed in a holistic manner to derive meaningful marketing intelligence about that person's overall banking behavior. Similarly, marketers in insurance companies can analyze policyholder's demographic information together with prior claim and premium information to derive meaningful marketing intelligence concerning a particular group of policyholders. The ability to do this, though, is contingent on capturing all behavior in a policy number, a number functioning much like the account ID in banks.

RETAIL CHALLENGES

For retailers, though, marketing is often not that simple, even with loyalty programs. For example, suppose retailer XYZ has a loyalty program including 200,000 customers. Customer transactions regarding the loyalty program are captured through a card the consumer presents at the point of purchase. Suppose a member goes to a store and uses the loyalty card to purchase a pair of tennis shoes at $100 because the salesperson prompted the customer to use the card by asking for it. And suppose that a week later, the same customer goes to a different store and decides to use his or her AMEX card and spend $200 on a jacket. Let's say the customer forgot to use the loyalty card because the salesperson did not remind him or her to use it or ask for it. The AMEX transaction, card number, amount, and date would be captured because this information has to be transmitted back to American Express for billing purposes. But how can retailer XYZ use this Amex information at the level of individual customers? The answer is that the retailer can't do that; only information from the transactions that can be attributed to the loyalty card can be used at an individual level.

How Can Nonindividual Data Be Used for Retailing?

But what about the Amex transactions or other transactions not involving loyalty cards? Is this information still useful? The answer is yes, but the information must be viewed from a different perspective. In the Amex case, the member's Amex behavior can be grouped with other Amex transactions at the store level. Store statistics such as the percentage of all transactions involving Amex cards or the percentage of the total amount due to Amex can easily be created and can be produced based on a certain time period. This approach can also be used for other types of payment vehicles, such as MasterCard, Visa, and even cash. For cash, there may be no electronic recording of the transaction as they involve the exchange of currency. Despite this information not being recorded, businesses can still approximate the level of cash transactions through default because for each store the total number and amount of all transactions and those of credit card transactions are known. The number of total cash transactions for a given store in this case is simply the difference between the total transactions and the total noncash transactions.

With this information available at the store level, stores that generate high value are also identifiable. To calculate this number, analysts would not focus on the size of the store as the main criterion of value. Obviously, a metric that calculates the average sale value per square foot would be a better representation of a store's value. Having identified stores that are of high value, analysts can then determine whether payment type is a key determinant of a store's high value.

Using Stats Can data, analysts can define trading areas around these high-value stores, and Stats Can demographic characteristics that pertain to these areas can be combined with other data. With analytics marketers can then identify the key demographic characteristics that differentiate a high-value store from a regular store. This information could be used in a number of ways:

- Targeting of prospects who live in these high-value trading areas or areas that look like high-value trading areas
- Opening of new stores in trading areas that look like high-value store trading areas

- Potential store closings based on the fact that they are located in low-value trading areas
- In-store programs based on the key demographic characteristics of customers who live in the trading area of a high-value store

Despite the information being available at an aggregate level only, marketers can use this information to help develop store programs and strategies and acquisition programs to attract new customers.

How Can Loyalty Card Information Be Used?

As previously stated, information accumulated in a loyalty program can be used at the individual level with more robust techniques, such as predictive modeling, used for this group of customers. Cross-selling and upselling programs as well as retention programs can all be based on customized predictive models to identify key customer segments for a given customer loyalty program.

For these types of analytics an important consideration is whether loyalty card behavior represents the customer's total purchase behavior and whether the difference between total transaction behavior and total loyalty card transaction behavior contributes to significantly different findings as compared to those based only on loyalty card behavior. Market research can help better understand the impact of this gap. Yet, data miners are very consistent in expressing the caveat that what consumers say in a market research survey is often very different from what they will do, the behavior that is captured in a database. Given this dilemma, the challenge is to apply these data mining techniques to loyalty card customer behavior while keeping this limitation in mind. Rather than spending a lot of time developing strategies to offset this gap, marketers have realized that they are better off dealing with what is known. In this case, marketers should focus on the loyalty behavior of a given customer. In other words, even though a client may in fact be the best customer of retailer XYZ, if that individual is a low-value loyalty card customer, loyalty marketing program and other promotions should be geared to that individual as a low-value customer in that context. The customer's overall high-value behavior is simply irrelevant because it cannot be used in a CRM program.

USING INFORMATION TO DEVELOP PRODUCT STRATEGIES

Retailing is focused on product movement. Retailers' primary objectives are quickly turning over inventory and establishing appropriate inventory control procedures. Data mining can help achieve these objectives and give retailers insights to use in developing product strategies for a given store. What kind of information is available to the data miner and how is it used? Information at an individual level represents the optimum scenario for data mining and analytics since the use of statistics is better leveraged through this increased granularity of information. At the store level, information is gathered at an individual level, but the store record itself is at the transaction level rather than the customer level. In this exercise, the transaction becomes the focus of the analysis as retailers strive to better understand what occurs in the transaction event. Retailers will want to know what type of products is bought in a given transaction. More important, are there certain types of products that appear to be bought together? And do these product relationships differ by type of store?

Observations by staff members combined with store report statistics yield valuable information on what products are sold and, more important, on changes in products sold in a given time period. However, what if the store is selling hundreds of products, and hundreds of thousands of transactions are occurring every month? Can this information still be considered reliable? The answer is yes, but data mining helps to supplement this knowledge by uncovering patterns that might not be observed through anecdotal observations or standard store reports. Through rigorous analysis, data mining can identify unique product patterns in a given transaction. For instance, if product A is purchased, what might be the next most likely product purchased in the same transaction event? Using product affinity and basket analysis, retailers can develop product bundling strategies for a given store. These strategies might be simple, such as placing products with high affinity in close proximity to each other and offering in-store promotions collectively marketing a particular bundle of products. Or at the point of sale the cashier could point out another product to customers based on what they are already purchasing.

An important aspect of data mining success is treating information as a valuable corporate asset. This is certainly no less important for the retail sector. But this success can only be realized if retailers understand that information is not only an asset but also a critical competitive advantage.

With this mind-set, retailers can truly differentiate their products and services in the competitive marketplace.

CAMPAIGN MANAGEMENT SYSTEM

This book has certainly emphasized the importance of data as the foundation of success. But in many cases challenges exist in the way data is organized in an organization. In many cases, this applies to how customers are being promoted and also to how they interact with the organization. The development of campaign management systems provides solutions to improve the organization of promotion and consumer response information. In a particular case example, the client achieved relative proficiency in using predictive modeling techniques to better market its insurance products. The company, a retail organization, marketed a variety of insurance products sold by various suppliers but marketed based on the retailer's credit card customer database. This rich credit card customer database became very competitive property as each supplier wanted to market its products to the best names as determined by a predictive model. If no rules were in place, each supplier would ultimately be promoting to the same names to obtain the best results. Within a very short time, list fatigue as well as increasing customer complaints followed by attrition would be the ultimate consequences. In preventing this situation, the company instituted a set of arbitrary rules so that each supplier would be allowed to promote to a given name only once in any 12 months. However, was this restriction reasonable? Without any analysis, this question would remain unanswered. The retailer understood that to answer this question, a campaign management database would have to be developed. In other words, the organization had to record information at the customer level pertaining to frequency of promotion, timing of promotion, and type of promotion. In addition, the company had to keep track of the outcome of the promotion effort (i.e., did the customer respond or what was the nonresponse outcome, if any?). The challenge then was to build this campaign management database that would help to determine the right number of promotions, the timing of promotions, type of promotion, and the actual promotion/campaign outcome that optimized campaign performance.

The first priority was to conduct a preliminary needs analysis to obtain detailed knowledge on the current data environment. Having built models in the past, analysts knew that solid knowledge regarding the data environment was important. However, it was still important to understand how data was transferred between the retailer and its insurance partners. This meant that when a campaign occurred, the following had to be understood:

1. What are the steps or activities required to create lists after scoring the names with the various insurance models?
2. What data processing activities occurred with campaign feedback?
   - How are respondents tracked and how are they fed back into the retailer's database to determine purchase activity?
   - Is nonresponder activity captured and how is it recorded? This is particularly relevant for telemarketing programs where the telemarketer will record the nonresponder outcomes, such as wrong number, call again, do not call again, etc.
3. Is the campaign activity captured historically?

In points 1 and 2, the detailed investigation allowed the company to produce a project plan and data flow chart. More important, the plan outlined how data would flow into and out of the campaign management database. However, in point 3, analysts observed that no prior campaign activity was captured. Essentially, this database had to be created in order to capture this information. First, this required the identification of the appropriate information to be captured within the database. Second, how the information should be organized in the database had to be outlined. The major information components required to be captured in this database were as follows:

- Model behavior
- Promotion behavior
- Disposition behavior

With this information, analysts could determine how customer behavior was impacted by the following:

- Changing model behavior
- Recency, frequency, and type of promotion
- Changing customer disposition activity

Because speed of processing was not an issue, the organization's structure was very basic in that simple database tables were created to store this

Campaign	# of Leads	# of Sales	Total Cost	Cost/ Sale	Avg. Prem/ Cust./ Month	# of Months to B/E	
1	20,000	285	$32,000	$112	$2.10	53	Pre Modeling
2	20,000	303	$32,000	$106	$2.34	45	
3	40,000	1,134	$64,000	$56	$4.17	14	Modeling Only
4	30,000	1,029	$50,000	$49	$4.40	11	
5	30,000	1,084	$54,750	$51	$4.06	12	
6	15,000	806	$30,446	$38	$3.89	10	
7	15,000	757	$28,442	$38	$4.79	8	Modeling & Contact Management
8	15,000	727	$26,678	$37	$4.72	8	
9	15,000	690	$28,064	$41	$4.10	10	
10	15,000	725	$27,225	$38	$5.07	7	

Figure 24.1 Table Depicting Payback in Months After Adopting Certain Data Mining Techniques

information. Yet, although the design was simple, the value to the client was designing a process that allowed the retailer to achieve the business objective of being able to make better decisions based on a customer's prior campaign history. The results after nine months of this launch are shown in figure 24.1.

This chart in figure 24.1 indicates how the institution of this system improved overall results. Using the key metric of average premium per customer, a new metric can be created to determine how many months it would take for a marketing program to break even given this program's costs. Obviously, a program that generates few responders and a low premium would require a longer time period before the program would pay for itself. In this chart, the time to pay back is approximately 50 months prior to any modeling technology being employed. Once modeling was introduced, this time was dramatically reduced to approximately 12 months. Deployment of modeling yielded the largest benefit simply because no analytics had been done prior to this activity. Future analytics work will yield benefits, but these won't be as significant since the reference point is now modeled activity as opposed to random activity. Introduction of the campaign management system did reduce the time to payback from 12 months to approximately 8 months. As the retailer uses this system overtime, more intelligence will be gained that will provide further insight in how to best reduce this metric.

CHAPTER 25

BUSINESS-TO-BUSINESS EXAMPLE

THIS CASE DIFFERS FROM THE OTHER EXAMPLE DISCUSSED IN THE previous chapter in that the direct customer is a small or medium-sized business rather than an individual. Once again, the company in question, a courier company, wanted to refocus its marketing efforts and essentially initiate a CRM program. As discussed previously in this book, management wanted to identify the company's best customers and profile them to better understand them. Yet, in creating or establishing metrics to identify the best customers, only purchase behavior was considered to keep things simple. However, this seemingly simple metric was not so easy to determine. Purchase behavior was often very sporadic and infrequent. The first challenge, then, was to determine the appropriate time frame for a small to medium-sized company in purchasing this company's courier services. In other words, do active customers make regular purchases in intervals of 1 month, 3 months, 6 months, etc.? Once this timeframe was determined, the company could then use the purchase metric as a proxy for value, albeit a simplistic measure.

A further component to the analytical exercise—in addition to segmenting customers based on value so as to identify the best customers—was segmenting customers based on recent behavior. Were customers increasing or decreasing their purchase patterns or were they perhaps defecting? For segmentation based on both value and behavior we adopted the VBS approach or Value/Behavior Segmentation. The first challenge, though, was to determine the appropriate time window for purchase behaviors. The approach here was to consider purchase patterns within different time frames and see if these patterns were stable or volatile. These time windows were then split equally into before and after periods to capture changing behavior in the periods before and after the time window examined. For example, with a one-month purchase window, analysts would look at purchase activity in one month and compare it to activity in the following month. Similarly, with a three-month window, analysts would examine activity within a three-month period and then compare this to the activity to the subsequent three-month period. Several options were considered in defining the purchase window. Several purchase window options were examined:

1 month
2 months,
3 months

Then customers were categorized into the following five groups based on these purchase patterns:

- Growers, that is, customers who were in the top 10% of purchase change behavior based on purchase amount and volume in the before and after time periods
- Decliners, that is, customers in the bottom 10% of purchase change behavior based on purchase amount and volume in the before and after time periods
- Stable, that is, customers not in the first two groups but with activity in both the before and after time periods
- Reactivations, that is, customers with no activity in the before period but activity in the after period
- Defectors, that is, customers with activity in the before period but no activity in the after period

The table in figure 25.1 table was produced to present our results to the business people.

This report looks at the three different options (1 month as the purchase window, 2 months as the purchase window, or 3 months as the purchase window) over a 12-month period and demonstrates how the time windows were evaluated. This report only takes into account 10 records for the sake of simplicity; in actuality, hundreds of records were assessed. Although this illustration visually identified the appropriate purchase time window, statistics were also employed to determine when the segment definition becomes constant (i.e., variance analysis) as the time period is increased.

	Month 1	Month 2	Month 3	Month 4	Month 5	Month 6	Option 1 1 month	Option 2 2 months	Option 3 3 months
Record1	$100	$0	$0	$200	$0	$0	inactive	defector	grower
Record2	$300	$0	$200	$150	$0	$75	reactivator	decliner	decliner
Record3	$500	$400	$300	$0	$0	$0	inactive	defector	defector
Record4	$0	$200	$0	$0	$50	$0	defector	reactivator	decliner
Record5	$0	$0	$0	$50	$75	$100	stable	grower	reactivator
Record6	$700	$600	$500	$400	$300	$200	stable	decliner	decliner
Record7	$0	$0	$400	$0	$300	$0	defector	decliner	stable
Record8	$0	$600	$0	$400	$0	$500	reactivator	stable	stable
Record9	$0	$0	$200	$0	$0	$400	reactivator	grower	stable
Record10	$450	$300	$250	$150	$100	$0	defector	decliner	decliner

Figure 25.1 Examining Different Pre and Post Purchase Window Options to Determine Appropriate Purchase Window

* Determined that 3 months was the optimum purchase window

In this report, the 1-month option demonstrates the appropriate segment for the given customer (small to medium-sized business). Extending the purchase window to 2 months, the segment definitions change quite a bit for the same customer and continue to change when the 3-month purchase window option is used. With the 4-month and 5-month options (not shown here) for the same records, the segment definitions become more stable and consistent with what the 3-month option showed. The implication here is that the appropriate window is 3 months as the segment definitions do not change when longer purchase periods are considered. Accordingly, all the behavioral segments were determined using 3 months as the normal purchase window for this company's customers.

The value segmentation was the next part of this exercise. The courier's business-to-business customers were segmented based on purchase volume in the past 12 months. The chart provided in figure 25.2 shows the rank order of these active business-to-business customers based on annual purchase revenue. As the chart shows, the top 20% of customers make up the high-value segment and account for average annual sales of $770, making up 68% of that courier company's total sales. In contrast, the low-value segment comprises most of the customer base (55%) and has average annual sales of $25, contributing only 7% of that courier company's total sales.

After identification of the value and behavioral segments, the dollar opportunity for each segment was calculated by multiplying the migration rate by the dollar value of that segment. Using the business knowledge

	Average Annual Sales	% of all sales captured by segment
High Value (Top 20%)	$770	68%
Medium Value (20%-45%)	$230	25%
Low Value (45%-100%)	$25	7%

Figure 25.2 Value Segmentation for B to B

Segment	$ value	Migration Rate	$ opportunity
High Value Grower	$800	1%	8
Medium Value Grower	$350	10%	35

Figure 25.3 Example of $ Opportunity for 2 Segments in B to B

of the courier company, we were able to infer a migration rate for each segment, and this allowed us to estimate the dollar opportunity for each value/behavioral segment, as is shown in the example of two segments in figure 25.3.

Based on these two segments, medium-value growers represented a better marketing opportunity than high-value growers. This approach, once adopted for all segments, allowed us to rank order all value behavior segments based on this dollar opportunity, and this yielded insight on how best to allocate the marketing budget.

This page intentionally left blank

CHAPTER 26

FINANCIAL INSTITUTION CASE STUDY

BACKGROUND

A financial institution used business rules to target regular credit card customers to become gold card customers. Specifically, the company used tenure and region of the country as the key segmentation criterion in selecting names. Results of using these criteria as list selection rules revealed that the targeted names barely outperformed the random lists. More important, the cost per new gold card acquired was continuing to increase from campaign to campaign. This was a problem because the company's objective was to significantly increase the number of gold card customers.

Given these facts, the organization required a solution that would optimize the number of gold cards and do so in a cost-effective manner. The financial institution identified the need for a tool, such as a predictive model, since that would have the following advantages over the existing list selection criteria:

- It could be applied to names beyond the boundaries of the list selection criteria, such as tenure and region of country
- It could consider variables and other characteristics beyond tenure and region of country that could achieve the objective of optimizing the response rate

The challenge was to develop a predictive model that optimized the response rate.

DATA CHALLENGE

Ideally, the best environment in developing a predictive model would be a random sample that has been promoted to in a previous campaign. In this particular situation, a campaign that occurred several months earlier had a random sample of 40,000 names. But this random sample was restricted to a group of customers who were already somewhat engaged with the company because they were already using the card on a monthly level. In this case study, this random sample was mailed to a group of names of people who had spent over $1,200 a year or an average of $100.00 per month. The response information from this campaign was captured with the response vehicle (application) containing the account number of the regular card. With the account number, this information could be matched back to the mail file in order to identify responders and non-responders. Both nonresponders and responders were then matched to the following files by account number in order to append the following information:

- Name and address file containing basic biographical information, such as age, income, household size, etc.
- Credit history containing information about spending and delinquency over a given period of time
- Banking file containing the customer relationships, such as loans, deposit/checking accounts, RRSPs (Registered Retirement Savings Program), etc.

As previously discussed, when building predictive models, the analytical file should be crafted into a pre/post window. If the campaign was dropped in July, the only information after July that is relevant for model-building is whether or not the prospects responded. All other information, which is being considered as potential model variables, must be viewed prior to July.

There were no challenges in matching or linking files together since the matching logic used an account identification number. Yet, the real challenge was the richness of data and creating all the potential model variables. There was no data mart or marketing warehouse at the time of the model-building exercise. All information was contained in legacy systems, and this implied that the data was contained in raw transaction and billing systems. The task of organizing and summarizing the data into meaningful information was very onerous and the most time-consuming portion of this project due to the volume of data as well as the number of variables (300+) that were created in this process. In addition, Statistics Canada data of over 100 variables was appended to these records by postal code. At the end of this process, the analytical file had 40,000 records with the dependent variable being response, and over 400 variables would be analyzed as potential predictor variables of the model. Using the modeling methodology discussed previously in this book (chapter 12), a solution was obtained that contained the following profile of a gold card responder:

- Higher behavior score
- Less likely to have an RRSP
- Likes to travel
- Does not live in Quebec

In addition to the targeting capability, the profile variables provided insight on various communication approaches (less likely to have RRSP and like to travel).

Using a cutoff of 40% as the cutoff or 200000 names, the model yielded a benefit of $96 million, which represented the opportunity cost of promoting to additional names without modeling to achieve the same level of responders as was achieved with modeling.

Figure 26.1 was set up to test the model's effectiveness and also to test different communication pieces. Clearly, most of the names being selected are in the top two quintiles along with the test communication because the expectation is that these cells will deliver the best performance.

% of File (Ranked Model Score)	# of Names Mailed	Test Cell	Control Cell
0–20%	100,000	1–95,000	1–5,000
20–40%	100,000	2–95,000	2–5,000
40–60%	100,000	3–5,000	3–5,000
60–80%	100,000	4–5,000	4–5,000
80–100%	100,000	5–5,000	5–5,000

Figure 26.1 Marketing Matrix after Implementation of Model

The actual results of this campaign made themselves felt quite early in the campaign due to increased traffic in the operations department where staff had to process an increasing number of applications. The final results of this campaign were as show in figure 26.2.

Campaign Test Results	Actual Response Rate
Modelled Control Names (@ 40% cut-off)	1.87%
Random Control Names (Combine all 5 cells)	1.10%
Actual $ Benefits of Model due to Saved Mailing Costs (@ $0.80 per mail piece)	$112,000

Figure 26.2 Actual Test Results

CHAPTER 27

USING MARKETING ANALYTICS IN THE TRAVEL/ENTERTAINMENT INDUSTRY

THE GROWING IMPORTANCE OF ANALYTICS AND DATA MINING HAS LED to the development of solutions in industry sectors such as travel and entertainment. Let's look at sports as one example of entertainment. Here, a flagship team like the Toronto Maple Leafs or New York Yankees does not need to be very effective to achieve its goal of putting more fans in stadium seats in a cost-effective manner. The reality for these organizations is that the product they offer is highly price inelastic—that is, the price charged for the product can easily be increased or decreased with minimal impact on overall ticket demand. The use of analytics and data mining in this context is meaningless.

However, for most sports organizations, putting more fans into seats in a cost-effective manner is a primary objective. Historically, sports marketers have attempted to increase attendance and grow ticket revenue by simply allocating more marketing dollars to mass advertising. But this has not always achieved the objective of increasing ticket sales in a cost-effective manner. Sports marketers need to understand that all fans are not created equal, an insight reminiscent of the standard CRM slogan that "not all customers are created equal." The capability to take customer differences into account is based on data and depends on whether information is being captured and stored somewhere. Unless someone at the receiving end of a ticket purchase transaction is capturing information about the ticket buyer, the activity or event becomes unavailable for any future analytics. It is important to understand this because a significant portion of a team's ticket revenues could come from unrecorded events, such as when tickets are purchased with cash at an arena or ball park. This limitation of data capture at the individual level obviously represents a barrier to many sports organizations that are considering the use of analytics as part of their marketing efforts. To pursue analytics in a more meaningful way, sports organizations must think of alternative ways of capturing customers' transaction behavior for future analytical purposes.

Digital technology is making it easier to capture purchase behavior because the payment process is now facilitated through the option of online purchases. With creativity, marketers can expand information capture beyond the ticket purchase to include concessions and merchandise. Through increased ability to capture data at the individual level, marketers can develop better programs to encourage incremental spend in ticket transactions and other transactions not involving tickets. This growing trend of online purchases has increased marketers' ability to capture data about individuals and, more important, to use this data and information as a way to differentiate between customers.

After sports, let's look at the hotel sector, which is another significant part of the travel/entertainment industry. Historically speaking, the needs of a hotel customer were quite narrow with customer mobility being a key limiting barrier. In the past, the population was arguably more homogeneous with many similar needs. In today's more complex world, customer needs are much more heterogeneous. What makes this challenge even more daunting is the fact that our constantly busy society leaves less time to explore these varying customer needs. Yet, even with this reduced time, consumer expectations are not diminishing but are in fact growing as technology has provided organizations and businesses with the basic capability of doing more with less.

In the travel industry, customers have always considered their time at a hotel as an experience rather than as just a visit. Activities such as fine dining, nightly entertainment, spas, and corporate seminars/meetings nurture this concept of "customer experience." It is easy to see that this range of activities is going to have varying levels of appeal to a given clientele. Obviously, the role of data mining and analytics can be quite significant in helping marketers to better understand these clients' different needs.

The first task might be to conduct a basic customer value exercise so as to identify the best customers. However, as with many analytical exercises, seasonality must be considered, and it is arguably even more significant for the hotel industry than for others. For many projects, marketers make the assumption that the rank ordering of customers is unaffected by seasonality. In other words, customers may spend more at different times of the year, but their spending relative to each other remains unchanged. Most analysts would agree that for the travel industry the issue of seasonality potentially has a significant impact on travel behavior. For example, one person may spend $1,000 annually as a casual traveler over the course of a year and is thus considered an average customer. Another person may spend $1,000 annually but spend the entire amount on a tennis package for one week in August. Should this second customer be in the same category as the first person because the annual spend ($1,000) was the same? Obviously, the answer is no. These two travelers are very different types of customers. Seasonality is also significant when conducting any analytics exercise particularly if consideration is given to the many hotels that will offer tennis and golf packages in the summer and ski packages in the winter. In addition to seasonality, various other services may appeal to different groups of customers and thus affect travel behavior. Fine dining and theater may appeal to one group of customers while spas and perhaps valet services appeal to another group. With varying interests among the travel clientele, a segmentation exercise categorizing the customers into clusters would be useful in identifying different customer segments' needs and interests. Travel industry experts surely understand that there are distinct customer segments. Based on the data being captured on these customers, businesses can then apply some science to identify distinct segments in their customer base. Specific marketing programs and initiatives can then be designed for each of these segments.

This page intentionally left blank

CHAPTER 28

DATA MINING FOR CUSTOMER LOYALTY: A PERSPECTIVE

THE DEFINITION OF CUSTOMER LOYALTY HAS LONG BEEN DEBATED IN THE marketing community. At issue is not so much the semantic definition of customer loyalty—most marketers would agree that loyalty is shown by a strong affinity or attachment to a given company's products or services. The differences arise when marketers attempt to measure or evaluate customer loyalty.

The implications here are that marketers must first *define* metrics and measures of customer loyalty. Once metrics are defined, it must then be determined whether the data is readily available in the current information environment for the creation of these measures.

IS RFM AN ACCURATE MEASURE OF CUSTOMER LOYALTY?

The most common approach taken to measure customer loyalty is to look at prior purchase behavior. One very common, quick, and pragmatic way of doing this is through the development of an RFM measure where:

R = recency of last purchase

F = frequency of purchase within a given period of time

M = monetary value of purchase(s)

By combining these measures, RFM indexes or scores can then be produced to generate a relative measure of loyalty for each customer. The higher the score, the greater the relative loyalty of the customer.

Many loyalty experts and pundits will debate this idea as being overly simplistic and not really capable of capturing the true essence of customer loyalty. They argue that there are groups of customers with high RFM scores who are not really loyal and will randomly switch companies or brands when provided with a better offer. Their so-called loyalty is not really due to any attachment or affinity to the company or brand; rather, they consider switching to another organization for the same products and services simply involves too much effort and/or risk. This weak affinity bond may be broken quickly when this type of customer is presented with an easy, risk-free offer.

Does it matter that some customers scoring high on the RFM index are not really that loyal? Does loyalty really matter if these customers are (in most cases) the most profitable customers? Usually, the value these customers deliver to the organization will result in the creation of "Best Customer" marketing programs that target this group regardless of the relative strength of these customers' loyalty.

USING MARKETING PROGRAMS TO BOOST CUSTOMER LOYALTY IN THE SHORT TERM

Whether a group scores high on the RFM index or has high value (or both), the next challenge is to acquire basic information about the effects of marketing activities boosting customers' attachment/affinity to an organization's products/services so that customer loyalty/value can be increased. One way of evaluating the effects of these programs is to look at customer behavior that is truly *incremental* when compared to past performance.

Changes in customer behavior can be defined in three ways:

- Lift
- Shift
- Retention

Lift refers to increased usage of a product or service; *shift* refers to the acquisition of new customers for a product or service; *retention* means that a customer's current activity level with a given product or service is maintained. In each case, behavior is evaluated to determine whether it is truly incremental as a result of a specific marketing activity. The key in identifying these behaviors (lift, shift, and retention) as incremental is the creation of an appropriate strategy for testing marketing programs. Specific groups of customers are segmented randomly into test cells and control cells; test cells represent the group being exposed to the marketing activity, and control cells consist of the group not exposed to the marketing activity. After the execution of the marketing activity, differences in performance between the two groups can be compared, and this ultimately determines whether an incremental effect was created by the marketing stimulus.

The use of RFM allows analysts to identify groups that may have a better ROI potential. The use of marketing activities with highly profitable groups allows for a proactive impact on the behavior of specific groups of customers. Successful marketing program testing allows analysts to definitively determine the incremental changes in customer behavior resulting from marketing programs.

Defining Customer Loyalty over the Long Term

Whereas previous purchase behavior may be an adequate measure of customer loyalty in the short term, there are other behavioral indices not involving purchase that must be considered when creating measures of long-term loyalty. Three customer behaviors that might be considered key information in identifying long-term loyalty include:

- Overall marketing response
- Customer inquiries
- Customer complaints

Overall marketing response refers to a customer's total number of responses to all the marketing campaigns of a given company. A customer's ongoing willingness to respond to campaigns over time could certainly be construed as measure of that customer's level of engagement with the company. These responses could be related to purchase and also to other activities.

Customer inquiries generally are positive interactions as they reflect the customer's desire to obtain more information from the company. In this activity there is an implied sense of trust between the customer and the company. This trust factor can grow only as long as the outcome of each inquiry is positive in the mind of the customer. Conversely, *customer complaints* are negative interactions between a customer and a company. Erosion of trust between the customer and company may occur with each complaint.

Database marketers want to identify these traits or characteristics from the information recorded in the database. Marketing response can be captured when an organization has a campaign management system in place for its marketing campaigns. With such a system, campaign information and responses to those campaigns of individual customers can be captured, and this allows marketers to identify the overall marketing response history for each customer. Another rich source of customer inquiry information is a company's website. By examining web logs, it is possible to identify the page views of customers. The page views in most cases represent basic inquiries for information. More page views of a company's website by customers mean that customers are generating more inquiries about the company's products and services, and arguably this indicates that they have a heightened interest in the company. Meanwhile in the telemarketing world, both outbound and inbound calls deal with customers who often inquire about other products and services beyond just the expression of complaints.

How is this information organized into meaningful loyalty measures? Internet page views and marketing response history (campaign management system) are well organized as the data is located in structured data fields. The analyst can simply go to the appropriate field whether it is a page request field on the web log file or the marketing response flag related to some specific campaign. In both cases, this information can be summarized from each of these fields to obtain specific views of customer activity or engagement. However, in the case of a telephone conversation, the information of interest is in the dialogue between the sales or customer service rep and the customer; that is, the information is present in an unstructured format. Unstructured data means that it is not possible to point to one specific field in the analysis; instead, the entire text of the dialogue must be reviewed to extract meaningful information. The concept of extracting meaningful information from unstructured data has become a primary focus of data mining research today and is commonly referred to as *text mining*. This will be discussed in the next chapter. Through this discipline, conversations, comments, and any other type of text can now be mined to extract meaningful information. In the case of extracting meaningful measures of customer loyalty, text mining can be used to identify customer complaints or customer interest levels based on the conversations or comments between a customer and a sales or customer service representative.

The use of campaign management systems, the Internet, and text mining provide new ways of creating customer loyalty measures beyond basic, short-term purchase behavior. Taking a longer term perspective on the development of customer loyalty measurement ensures that broader customer engagement and attachment behaviors are incorporated into these measures. Marketers can then be equipped with tools they can use in executing initiatives that increase customer engagement and drive long-term customer profitability.

CHAPTER 29

TEXT MINING: THE NEW DATA MINING FRONTIER

DATA MINERS AND ANALYSTS ARE USED TO DEALING WITH DATA IN structured fields, such as rows and columns. Data, such as age, income, purchase spending, and purchase dates, are typically found in specific columns or fields. Advances in data mining technology now allow analysis of unstructured text data. In other words, information pertaining to language and communication can now be analyzed. This emerging discipline is commonly referred to as text mining or text analytics. One question that may immediately come to mind is how this differs from search engine technology. In search engine technology, users list key words or phrases that are then analyzed to produce a list of articles, themes, or topics supposedly related to the key words. Text mining also analyzes text, but users do not enter specific key words or phrases; rather, users do not know what they are looking for. A body of text or unstructured data, called a corpus, is presented to the analyst. With text mining tools the analyst then identifies patterns or themes in this corpus. For example, XYZ company may want to use some market research analysis of its customer base to better understand why the number of complaints has increased. Text mining could be employed here to extract all the customer e-mails of the past several months. By analyzing all these emails, the analyst may uncover themes and/or discussion patterns that could help identify specific reasons for the increasing customer complaints.

Another good example is product development. Market research is often used to help in this area. For example, text mining technology can help analyze call logs of inbound telephone calls of customers to identify call patterns related to specific service needs. This information could then be used along with market research to help develop more appropriate products and services.

TRADITIONAL DATA MINING

Most experienced data mining and analytics practitioners have valuable expertise in mining data in the more traditional manner where data occurs within a structured format. A structured format means that the actual records themselves represent rows while the information that is used for data mining resides in columns.

Records represent the level of detail on how information is being captured. For example, records can be at the customer level, transaction level, and promotion level, or at any level where information is being captured. Meanwhile, columns can be thought of as variables that depict a certain piece of information pertaining to that record; examples of structured data formats are shown in figure 29.1.

In the example in figure 29.1, there is a customer file and a transaction file. In the customer file, individual customers are represented in the rows, and customer number, household size, postal code, and income represent the variables and are shown in columns. In the transaction file, individual transactions are represented in the rows, and transaction number, date, amount, and product type represent the variables columns. This structured

Customer Table

Customer No.	Household	Postal Code	Income
1	3	L1A3V1	$125,000
2	2	M5S2G1	$30,000
3	1	H4B2E5	$40,000

Transaction Table

Transaction N°	Date	Amount	Product Type
1	July 15/2009	$100.00	A
2	Oct 1/2009	$75.00	A
3	Sept 15/2009	$200.00	C

Figure 29.1 Example of Structured Data (Customer Table and Transaction Table)

approach provides great flexibility when it comes to data manipulation or the ability to derive new information from source information. For data miners, this is a critical capability, and in this book we have discussed at length how most of the information in a data mining project is derived.

EXAMPLES OF TEXT MINING

As we have discussed, in text mining users do not know initially what they are looking for, but text mining is a way to find patterns or trends in the unstructured content. In addition to marketing applications, which will be discussed later, text mining has many other business applications. Document categorization represents one significant application; here, historically much manual intervention was required to group documents into meaningful categories, and much of this work is now automated. However, human intervention is still needed to check the accuracy of the "automated process."

Text mining is similar to the standard discipline of data mining as a process of knowledge discovery. In both cases, users do not know what they are looking for and mountains of data are analyzed to provide insights. Unstructured data is textual data such as e-mail messages, phone conversations, open-ended responses to surveys, and conversations in various social media applications. This is an important consideration given the channels that marketers can now utilize in reaching out to customers. A very simple example of what an unstructured format might look like is shown in figure 29.2.

Customer Nº	E-mail
1	I really like the RRSP product and will continue to invest every year. However, level of service is sub-standard and I will be looking at other companies. But it will be difficult since I have many products with this institution.
2	I wish this company would be more proactive in offering me products and services that truly meet my needs. The customer service people are great and are simply doing their job. But the company is clearly not thinking of my real needs.
3	The level of service is outstanding and I am extremely interested in hearing about all your products and services. Could you send me more information on them?

Figure 29.2 Example of E-mail Unstructured Data

In this example, three customers have each sent one e-mail message to the company. The data mining approach here would be to analyze the information listed under the column heading “E-mail.” Here, analysts would be left with the task of trying to derive information from the e-mail column. How can they do this? They can create binary or yes/no variables based on some condition, create change variables, or mathematically derived variables, such as mean, standard deviation, median, etc. For example, in the structured world, based on the postal code field of where a person lives, a yes/no variable is easily created based on whether or not the person lives in Quebec. In the unstructured world, the challenge is how to derive new variables from a series of sentences. But the critical point to remember is that there is information in this text that can and should be used. The process of using this information by creating variables from unstructured data is not necessarily more complex than creating variables from structured data. It is just different as this discipline requires a different approach as well as different tools for the creation of meaningful information and insights. But what does this really mean?

The Text Mining Process

The first step in mining unstructured text data is to perform some cleanup or data hygiene. This means to eliminate text that really provides no information. Examples of this include punctuation marks, such as period, comma, etc., and prepositions, such as “the,” “of,” etc., and pronouns, such as “she,” “he,” etc. The example in figure 29.2 could be reduced and parsed with data hygiene techniques to what is shown in figure 29.3. That is, only the key information was extracted that is needed to develop some meaningful information.

Customer N°	E-mail
1	Really like RRSP product continue invest every year level service substandard looking other companies difficult have many products institution
2	Wish company proactive offering products services meet needs customer service people are great are doing job company clearly not thinking real needs
3	Level service is outstanding extremely interested hearing products services send information

Figure 29.3 Example of E-mail Text That Has Been Filtered for Data Hygiene

The next step is to perform a frequency distribution on all the keywords that remain after the data hygiene process. In this step the first glimpse of meaningful information is obtained in terms of how often certain words occur. After a simple frequency distribution, the data would look as shown in figure 29.4.

From the information, we can conclude that popular keywords being discussed are: product, services, company, needs, and level. However, this is still too early to give any meaningful concrete insight, and the tricky stage of the process has yet to be performed.

This third stage of the process is to find relationships between words and phrases that seem most prominent in the text. The linking of words and creation of common phrases represents the real selling feature for most companies selling text mining software. This is particularly true for text mining vendors selling to the government; they may be analyzing phone conversations in order to detect criminal activity. Two of the more common techniques that could be employed are correlation analysis (keyword analysis) and cluster or k-means analysis. Correlation analysis is used to find groups of words that appear together most frequently. Meanwhile, k-means or cluster analysis attempts to find phrases or groups of words that appear as common themes. At the same time, the clustering is used in an attempt to identify groups of themes that differ from each other. This is somewhat analogous to the traditional use of clustering, which aims to

Words	Frequency
Products	3
Service	3
Company	2
Level	2
Needs	2
Services	2
RRSP	1
Am	1
Are	1
Clearly	1
Companies	1
Etc.	

Figure 29.4 Example of Frequency Distribution on Keywords in e-mail

create customer segments where customers within a segment have similar behaviors and demographics and each customer differs from the others. In the three record cases in the example, some emerging themes might be:

1. Wide range of products and services
2. Need for information
3. Service levels

Sentiment Analysis

But even as the three common themes emerging from these records are identified, how customers feel or the sentiment they are trying to convey is not displayed among the three themes. The ability to represent the sentiment or emotion of the statement, often referred to as sentiment analysis, is an area of extensive research in text mining. In fact, such statements could be:

1. I am really going to like this company now that they have cut services as it takes me longer to get through to a live agent
2. I am really going to like this company now they have invested in new customer service software and hired another 1,000 agents

The first statement is sarcastic and expresses a negative sentiment while the second appears to be quite positive given the company's investment in resources related to services. Detecting the specific sentiment in each of these statements is fairly easy for the human brain based on understanding of language and tone. However, it is not easy to design software that can detect the tonality of these sentiments. In fact, two statements both stating "I really like your company," may express very different sentiments based on the tone in which they are uttered. Text mining tools cannot pick up differences regarding tonality, but an offshoot of text mining called speech analytics analyzes unstructured voice data and might have this capability. Analyzing sentiment in voice data is also extensively explored by speech analytics vendors.

The Marketing Interest

Marketers are interested in using text mining information because it represents another level of customer engagement that can be analyzed for marketing purposes. Despite the unstructured format, meaningful insights and intelligence can be gained from classifying customers based on theme categories evident in their individual discussions. In addition to theme classification, more advanced use of these tools can allow marketers to classify and segment customers based on sentiment (negative, positive, neutral).

Going Forward

Text mining and its application in some of the newer marketing channels, such as social media, is in its infancy. No one can really lay claim to being an expert in this area yet since any experience is going to be minimal at this point in time.

As with any new discipline, practitioners must to some extent rely on historical experience, and in this case, this means on their ability to deal with analytical challenges. Data miners and analysts in many cases have many years of experience in analytics related to structured data. Well-honed skills have been developed so that the measurement of ROI, for example, has become a core analytical imperative alongside the ability to better understand what impacts ROI. For many data analysts, these skills have helped to create an approach and process that will be useful in any analytical exercise. By leveraging this experience and expertise in the area of text mining, marketers can more fully exploit these new disciplines for the development of better marketing solutions that involve both structured and unstructured data.

This page intentionally left blank

CHAPTER 30

ANALYTICS AND DATA MINING FOR INSURANCE CLAIM RISK

THE PRIMARY ADVANTAGE OF USING ANALYTICS AND DATA MINING FOR P&C insurance is the ability to more effectively create a rating structure or to price premiums for a given policy. The challenge is how to build the best tools in this area, particularly how to develop multivariate analysis (MVA) tools. In other areas, such as marketing and credit card risk, the use of predictive analytics and associated multivariate analysis (MVA) tools has become commonplace. Why does this pose a challenge in the P&C industry; after all, actuaries usually have a strong background in mathematics? The answer lies not only in the mathematics being employed but in the lack of knowledge on how to use data in the right way to take full advantage of the techniques available. In order to fully leverage the results of any MVA tool, hundreds of thousands of individual policy records with several hundred variables per policy record must be created, and this is the core skill set of the data miner. The discipline of data mining is relatively new, and the training of actuaries is not fully developed in this area. Nevertheless, the academic training actuaries receive in mathematics and statistics can definitely be applied to data mining. However, the most important component of data mining, and arguably the one that is very resource-intensive, is the data environment. Unless the right data environment for this kind of analysis is created, the use of MVA techniques is not going to deliver superior results over the older pricing methods.

Most leading-edge organizations realize that to compensate for the lack of data-mining knowledge among actuaries data mining practitioners and actuaries must collaborate closely. Data mining practitioners will never have the knowledge and expertise actuaries have regarding insurance risk. This is particularly important if insurance rates need to be filed in accordance with government regulations. The actuaries' knowledge of mathematics and the insurance sector provides the credibility on which to establish a rating structure that is acceptable for a given market. However, the use of MVA tools, which are the traditional domain of data mining practitioners, allows organizations to further improve their rating structures. Knowledge of MVA techniques is not new to actuaries, but the creation of the necessary data environment is not part of their expertise. By combining actuaries' expertise with that of data mining practitioners, companies can produce MVA solutions that more fully leverage this data environment. These solutions can then be modified based on actuaries' knowledge of what is acceptable in a given market. In a sense, the final pricing or rating solution requires a team-based approach between data miners and actuaries. This means that data mining practitioners must better understand the filing mechanisms of actuaries, and actuaries must gain a better appreciation of the data environment and how critical it is to delivering an optimal MVA solution.

CHAPTER 31

FUTURE THOUGHTS: THE BIG DATA DISCUSSION AND THE KEY ROLES IN ANALYTICS

BIG DATA. HYPE AND MORE HYPE. BUT WHAT IS THE REALITY HERE? Is big data the real deal as many experts would have us believe? For seasoned data mining practitioners, big data has always been the norm rather than the exception. Building predictive models in the late eighties at American Express exposed practitioners to a million cardholder records with hundreds of millions of transaction records. The challenge, then, was to translate this raw source information into a meaningful analytical file that could be used as inputs for developing predictive models. At that time, marketing databases and data warehouses were new to practitioners. The use of crude tools, such as legacy mainframe systems, required a certain level of technical knowledge concerning the use of JCL (job control language) and the storage of data. Accessing data and data output on tape or disk required technical knowledge to optimize the use of data mining in this big data environment.

For instance, in creating analytical files, it was not unusual to have files with hundreds of variables. However, the analytical file was created through intense programming by practitioners. As with any programming routine, these programs had to be tested rigorously to ensure that the programming objectives of the analytical file are being achieved. To do this effectively, testing and quick feedback of the results represents the desired outcome. However, with a million records and hundreds of millions of transaction records, the data would be stored on tape, and the actual programming routine would be submitted as a batch job with the results being relayed the next day, not in seconds. For the most effective test of the programming routines used in the creation of the analytical file, practitioners would extract a small random portion of the raw data and store it on disk. By storing that data on a disk, practitioners could test their programs on the sample data and determine whether programming changes had to be made. With this approach, the analytical file could be created in a more timely manner even with the crude tools available back then. This approach to using random data would also have been employed in stage 3 of the data mining process, namely, the stage involving the use of statistics in the development of the model.

As the American Express example shows, the ingenuity with which practitioners dealt with information in a big data environment was key to building effective data mining tools in a timely manner. As the PC environment evolved and marketing databases and data warehouses were introduced, the improved technical infrastructure of big data significantly advanced the development of data mining as a discipline. Yet, despite technical improvements, data volumes have continued to explode due to the growing significance of CRM and the development of the Internet. The world of social media has further accelerated this explosive data growth. Are larger data volumes the real challenge for practitioners? The answer would be a resounding no. What are the challenges for practitioners in this big data world?

First, let's consider what the pundits say about big data and the three Vs, which are volume, variety, and velocity. As discussed, the volume component is definitely not a new phenomenon for practitioners. Yet, the other two Vs represent challenges for the practitioner. What do I mean here? Variety represents information and data that is arriving in different formats. This scenario is best exemplified in the unstructured and semistructured data that is the norm in the social media context. Blogs, videos, posts, text messages, e-mails, log files from web visits—these all are nonstructured data. Most practitioners are very comfortable in the big data volume environment as long as the data is structured. However, what happens when the data is unstructured? New tools, technologies, and skills are required in order to best optimize this information for data mining purposes.

The old approaches of extracting and mining information in structured rows and columns no longer apply. Tools that can extract pods of information rather than rows and columns and programming techniques that extract the critical information now allow analysts to work effectively in this new paradigm. Platforms such as Hadoop MapReduce allow organizations to process data more effectively through what is called distributed file processing. Then, tools, such as NoSQL, Python, Pig, Java, and many others provide the programming languages that allow meaningful information to be extracted. The critical task for data miners is to determine the best option to extract data and convert it to a variable or variables that will be meaningful for a given business objective.

The emphasis on big data and big data analytics has focused the data mining discussion on the importance of the data itself rather than on the mathematics. Once again, this scenario goes back to the basic four-step process of data mining; that is, the process begins with determining the data mining challenge or problem and then in a next step constructing the appropriate analytical file—and doing this in the big data environment. Most experienced practitioners welcome the discussion surrounding big data because it has increased awareness of data mining and, more important, of the process of creating the right data environment. New terms, such as data science, now testify to the importance of data in the data mining process. Data science means that practitioners undertake a rigorous and methodological approach in dealing with data. From a practitioner's standpoint, is any of this really new? Data as a discipline has always been the core foundation of developing in data mining.

The discussions around big data have also led to supposed new roles in the world of big data. It is amusing to see long-standing roles or functions being repositioned in the frenetic world of big data analytics. For long-time practitioners, a certain cynicism arises from role repositioning. Regression analysts of the early eighties became modeling analysts of the late eighties and then turned into the more commonly accepted term of data miners. Even disciplines such as direct marketing analytics are now encompassed under the general term big data analytics.

Let's take a closer look at the roles of data scientists and value architects (business users) who are involved in any data analytics project. Back in the eighties, these roles existed in many leading direct marketing companies. Direct marketing companies had key individual or modeling analysts, called data scientists by today, who were building direct mail response models. Yet, such models could only be used with the involvement of marketers, commonly called value architects today. This required a marriage of the two disciplines in the sense that data scientists had to have some knowledge of marketing, and the value architect had to have some knowledge of modeling. Yet, this so-called crossover knowledge would be irrelevant without deep domain knowledge being paramount for each role.

To better understand these two roles, it is useful to see what this means in direct marketing because that is where the history of building data mining solutions began. In the early days of analytics, the discipline was almost exclusively used in a tactical manner. Overall marketing strategy was determined by the key leaders in the marketing area. Specific tactics, such as marketing campaigns, were then designed and created to execute on this strategy. To ensure that these campaigns were as profitable as possible, analytics were conducted in order to optimize efficiencies, and these were measured in such performance indicators as reduced cost per order or reduced cost per customer saved.

In this very tactical exercise, these two roles (data scientist and value architect) are distinct yet complementary in achieving their objectives. The marketers usually have already developed the campaign or initiative and its accompanying budget. Optimizing budgets was the primary driver of analytics in the early days. That is, the analytics might provide insights into developing basic business rules that would reduce costs. For example, in acquiring new customers, the identification of key lists or particular Stats Can demographic segments was essential. At this level, marketers were simply trying to understand the key characteristics differentiating new customers from the general population. Over time the cost reduction from this learning would erode, which means that a more advanced analytical approach beyond basic business rules was needed. Marketers wanted to develop a model, and so they met with data scientists. The value architects or marketers would explain the objective of the campaign, and usually they would define the required modeling tool. Listening carefully to the marketers' campaign objectives, the modeling analysts would then think about the data to assess, for example, what data sources they had for building an acquisition or retention model. To develop a model, data scientists also must consider whether they can create the necessary information environment that will deliver meaningful inputs for a model and whether they have the data to create the objective function or target variable. More important, they must assess whether they have the software that allows them to create both the information and the analytical environment and leverage the various statistical routines.

In today's big data context, analytics is a much more collaborative process, and marketing and business strategies are much more data-driven. It is common for organizations to commence their foray into analytics with data discovery. Data discovery is a process to define a data strategy that addresses an organization's business needs. The collaborative approach to data discovery helps determine the necessary tactics for the execution of the strategy.

As the importance of big data has grown so has that of data scientists and value architects. With social media and text data being readily accessible, the ability to understand context and the business insights that might be meaningful will remain the purview of value architects. Meanwhile, the ability to manipulate and derive meaningful information, which is necessary for these insights, is the responsibility of data scientists. These roles will continue to grow in importance as big data and analytics become more entrenched in a company's corporate culture. This will result in more collaboration and a more symbiotic relationship between value architects and data scientists. Without this close collaboration between these professionals, companies will be at a disadvantage in this new age of information and big data.

Companies can gather data in ever increasing amounts, but how can they make sense of it all? The data is often disconnected, and as raw material or source information, this data is often meaningless unless it is converted into meaningful information.

Even structured raw data must be transformed into meaningful information. For example, postal code information by itself is useless for data mining purposes. However, grouping postal codes into categories, such as regions, suddenly transforms this data into more meaningful variables for analysis. The classic example of raw structured data converted into meaningful information is transaction data. Here, data miners or data scientists create and derive different kinds of transaction behavior variables. Raw transaction data can be summarized in countless permutations from basically three fields:

- Amount
- Date
- Type

In addition to nearly infinite summarization options, data scientists or data miners can also capture change. For example, with the date field, analysts can identify how behavior is changing over time and also link this change to the type of transaction.

That is, the ability to link data is essential. For example, information on online behavior, purchase/transaction information, campaign history, demographic behavior all needs to be linked in order to gain a more holistic view of that individual or customer. Here, technical skills are required for determining the relationship between files and tables and the way to link the tables. Often, analysts must construct the fields to link these tables, a task that requires technical expertise.

Transforming raw data into meaningful variables is complex and hard work. What are the key ingredients to achieve success? Going back again to our four-step data mining process, the first step is to identify the business problem. Then analysts must work their magic with the data. The essence of this magic derives from two areas:

- Domain knowledge of the business and of what is required for business to be successful
- Mapping the current data/information environment into meaningful variables derived on the basis of the above-mentioned problem or challenge

So far this discussion has been about what is important from the perspective of data scientists. Value architects or key business stakeholders have equally important perspectives. For example, data scientists focus primarily on the information environment, but value architects focus on the business domain. Their perspectives must overlap to some extent; one might argue that the most successful solutions arise where this overlap is pronounced while the specific core functional strengths of each area are maintained.

The growing importance of data as a key corporate asset make data scientists and value architects ever more important to organizations. People beginning a career in the field of data mining may eventually end up in the executive suite of organizations. It is my hope that this book enhances your journey into data mining and provides a reference point on key considerations.

INDEX

www.ingramcontent.com/pod-product-compliance
Lightning Source LLC
LaVergne TN
LVHW022201200926
848481LV00006B/1808

9781349487868